THE EFFECT OF HARVESTING AND FLOODING ON NUTRIENT CYCLING AND RETENTION IN *CYPERUS PAPYRUS* WETLANDS

Edwin M.A. Hes

Thesis committee

Promotor
Prof. Dr K.A. Irvine
Professor of Aquatic Ecosystems
IHE Delft Institute for Water Education / Professor of Aquatic Ecology and Water Quality
Management, Wageningen University & Research

Co-promotor
Dr A.A. van Dam
Associate Professor of Environmental Systems Analysis
IHE Delft Institute for Water Education

Other members
Prof. Dr W. de Vries, Wageningen University & Research
Prof. Dr J.T.A. Verhoeven, Utrecht University
Prof. Dr L. Triest, Vrije Universiteit Brussel, Belgium
Dr M. Saunders, University Dublin, Trinity College, Ireland

This research was conducted under the auspices of the Graduate School for Socio-Economic and Natural Sciences of the Environment (SENSE)

THE EFFECT OF HARVESTING AND FLOODING ON NUTRIENT CYCLING AND RETENTION IN *CYPERUS PAPYRUS* WETLANDS

Thesis
submitted in fulfilment of the requirements of
the Academic Board of Wageningen University and
the Academic Board of the IHE Delft Institute for Water Education
for the degree of doctor
to be defended in public
on Friday, 1 October 2021 at 4 p.m
in Wageningen, the Netherlands

by Edwin M.A. Hes

Born in Meppel, The Netherlands

CRC Press/Balkema is an imprint of the Taylor & Francis Group, an informa business

Published by:

CRC Press/Balkema

enquiries@taylorandfrancis.com

www.crcpress.com – www.taylorandfrancis.com

ISBN 978-1-032-19461-5 (Taylor & Francis Group)

ISBN 978-94-6395-884-4 (Wageningen University)

DOI: http://doi.org/10.18174/549700

Voor Koen, Nadia en Vivianne

Acknowledgements

A journey of learning has resulted in the thesis before you. I'm proud and happy that parts of this journey were shared with other learners. Therefore, I'd like to thank my fellow travellers: Niu Ren, Ruth Yatoi, Sadiki Laisser, Do Phuong Hien, Kuenzang Tshering, Hawa Nakitende, Aster Feyissa, Julius Kipkemboi, Jeroen de Klein and Jan Janse.

I would like to thank Erwin Ploeger, Erik de Ruyter van Steveninck, Huub Gijzen and Henk Lubberding for supporting and motivating me to continue my career at IHE as an academic. Without the encouragements of the four of you this project would never have started. A special thank you to my promotor, Ken Irvine, for being supportive, patient and available, for your valuable inputs and perhaps most of all for sharing your genuine passion for freshwater ecosystems. IHE is a diverse community in many ways and I would like to thank all of you that have taught me valuable lessons along the way. Perhaps the most valuable lessons I learned from Gretchen Gettel who showed me the art of asking questions, Klaas Schwartz for opening the door to the complex and scary world of social sciences, Jeltsje Kemerink for demonstrating that it is always worthwhile to keep fighting for what you believe is important, John Simaika for demonstrating that the tiny ones matter even more than I already knew, and Jan Herman Koster for being Jan Herman Koster. Finally, a big thank you to all the 'Masters' and 'wannabee Masters' for our shared interest in what is really important. I would also like to thank the "Dream-Team" members Lakshmi Charli Joseph and Ingrid Gevers for all our shared wetland adventures, for being two of my biggest teachers and for our unique friendship.

I'm grateful to Hennie, Arie, Yvonne, Rob, Annelies and Leo for forming and inspiring me in your own individual unique ways.

Thank you Luuk, Karin, Edwin, Arjan, Harrie, Erik, Frank, Lässlo, Almi and Jochem for being supportive, always! Marco, thank you for being my paranymph, a sounding board, for letting me win every now and then and for supporting the right team. Wim, also thank you for being my paranymph, for being the wonderful person that you are, for showing emotions, the Grutto, Ede Staal and so much more and for also supporting the right team.

Vivianne, it is not easy to capture in words how much and in how many ways you have contributed to this thesis. So, I will not even try and limit it to the example you set with your career and how you have built that by being so true to yourself and your values. Nadia en Koen, jullie blik op de wereld en de spiegel die jullie mij op geheel eigen wijze voorhouden hebben mij geholpen en geïnspireerd. Ik ben zo enorm trots op jullie!

Finally, I would like to thank my co-promotor Anne van Dam, without him this journey would not have been completed. Thank you for your expertise, for giving me confidence when I ran out, for the joint writing trips, for also being a member of the 'Dream-Team', for being there for so many learners, for being humble and for being a great friend.

SUMMARY

Cyperus papryus L. is a fast-growing sedge that can grow to lengths of 5 meter and dominates the vegetation in many wetlands from the Middle East, through eastern and central Africa to the south of that continent. Papyrus wetlands traditionally have provided important ecosystem services to millions of people through provisioning of water, fish, other foods and medicines, and materials for a wide range of uses such as construction, utensils and others. Papyrus wetlands are also important because of their regulating and cultural ecosystem services and their biodiversity. In the landscape they are buffers for water and nutrients, provide habitat for fish, birds and other wildlife, and have been an integral part of the livelihoods and culture of African wetland communities for ages. African wetlands are under pressure from human activities and economic development. As a result, the total area of these ecosystems is declining. With a growing African population and demand for food production, protecting wetlands and combining increased agricultural production with conservation of the ecological integrity of wetlands is urgent. Eutrophication is of growing concern in the region and papyrus wetlands are known to reduce nitrogen (N) and phosphorus (P) loads to downstream water bodies. An increased ability to quantify the role of wetlands in N and P retention and a better understanding of the role of the different processes contributing to N and P retention is needed. To improve understanding of the processes contributing to N and P retention in papyrus wetlands and to evaluate the effects of different management regimes (particularly changes in water regime, and vegetation harvesting), the main objective of this thesis is to develop a dynamic simulation model for nutrient cycling and retention in rooted papyrus wetlands. *Chapter 1* of this thesis presents the background on papyrus wetlands in Africa and their ecosystem services, and briefly introduces the thesis chapters and their objectives.

In *Chapter 2*, the effect of seasonal flooding and livelihood activities on retention of N and P in *Cyperus papyrus* wetlands was investigated, with a focus on the role of aboveground biomass. This was done in two wetland sites in East Africa under seasonally and permanently flooded conditions. Nyando Wetland in Kenya was under anthropogenic disturbance from agriculture and vegetation harvesting, whereas Mara Wetland in Tanzania was less disturbed. The growth of individual papyrus culms and their density was monitored for a period of three months in a replicated field experiment, with 1 m^2 quadrats in both the seasonally and permanently flooded zones. Maximum papyrus culm growth was described well (regressions for culm length with R^2-values from 0.70 to 0.99) by a logistic model, with culms growing faster (r-value of 0.081-0.097 d^{-1} versus 0.071-0.072 d^{-1}, respectively) but not taller in Nyando than in Mara. Maximum culm length was greater in permanently (413 in Nyando, 484 cm in Mara) than in seasonally flooded zones (287 and 464 cm, respectively). Total aboveground biomass was higher in Mara (4.4-7.5 kg m^{-2}) than in Nyando (1.4-2.5 kg m^{-2}). The amounts of N and P stored were higher in Mara than in Nyando. Based on these results, it was concluded that papyrus plants in disturbed sites show characteristics of r-selected species which leads to fast growth in a short-term response to a disturbance but lower biomass and nutrient storage in the longer term. In less disturbed sites, papyrus growth resembles K-selected species with slower growth rates initially but a higher maximum biomass and nutrient storage. These findings help to optimize management of nutrient retention in natural and constructed wetlands, for example by determining optimum harvesting strategies.

As a first step in developing a model to quantify nutrient retention and the impact of harvesting in *Cyperus papyrus* wetlands, in *Chapter 3* a review of wetland models is presented. The chapter explores the outlines of a model with the main objective to assess N and P retention in papyrus wetlands, and how these processes are affected by both natural and anthropogenic drivers. A review of the scientific literature on wetland models resulted in 75 publications that were divided in four categories: hydrological models (17), biogeochemical models (20), vegetation models (28) and integrated models (10). Focusing on the water, nutrient and plant processes at the wetland scale, the review then concentrated on the publications on biogeochemical and vegetation models from which ten models were selected that all included N or P processes. These ten models were used to identify in more detail the required inputs for a papyrus model, such as water inflow, rainfall, evapotranspiration, solar radiation, nutrients (N and P), biomass growth characteristics, biomass harvesting, and soil porosity. The potential outputs of a model were also listed: water volume, inundation levels, above- and belowground biomass, N and P retention, N and P outputs in runoff and seepage water. A conceptual model was then proposed for 1 m^2 of papyrus wetlands with these in- and outputs and a simple hydrological component as forcing factor. Because of the characteristics of papyrus and the role of rhizomes in storage of N and P and reproduction, the distinction between above- and belowground biomass and the interaction between N and P as potentially limiting nutrients should be included in the model. It was concluded that a total biomass approach was most suitable for the papyrus model, partly because data limitations might restrict modelling of individual culm growth.

Chapter 4 then presents the first version of a simulation model for N cycling in natural rooted papyrus wetlands in East Africa. The model consisted of sub-models for the permanently and seasonally flooded zones and was based on data from a papyrus wetland in Naivasha, Kenya. In each zone, water, N and carbon flows were calculated based on descriptions of hydrological processes, such as river flow, lake level, precipitation, evaporation; and ecological processes, including photosynthesis, N uptake, mineralisation, and nitrification. Literature data from the extensive research on the papyrus wetlands at Lake Naivasha (starting in the 1970s) in Kenya, complemented with data from publications on other papyrus wetlands were used for parameterization and calibration. The model consisted of 35 state parameters and 143 rate variables and was implemented using the Stella software for system dynamics. Based on a comparison with values from the literature, the model simulated realistic concentrations of dissolved N and papyrus biomass density. Daily harvesting up to about 84 g dry matter $m^{-2} d^{-1}$ (in the seasonally flooded zone) and 60 g dry matter $m^{-2} d^{-1}$ (in the permanently flooded zone) reduced the aboveground biomass and increased N retention (expressed as (N_{inflow}-$N_{outflow}$)/N_{inflow} * 100%) to 38% (seasonally flooded zone) and 50% (permanently flooded zone). A further increase in daily harvesting resulted in collapse of the aboveground biomass.

Papyrus biomass, however, recovered fully from annual harvesting of up to 100% of the biomass. The model showed that papyrus re-growth after harvesting in the permanently flooded zone was N-limited because N could not accumulate to replenish the N taken up by plant growth as water flowed out to the lake.

In *Chapter 5*, the model (now called 'Papyrus Simulator') was developed further and a sensitivity analysis was performed. A phosphorus (P) section was added, and the model was generalized by merging the original seasonal and permanent wetland zones (based on Lake Naivasha) into one wetland zone that can be inundated based on the prevailing conditions of stream and overland flows, lake inflow (if applicable), and precipitation. The main output variables of the model (aboveground and belowground biomass, N and P concentrations in biomass and water, net biomass production) were compared with data from a review of 22 papyrus wetland field studies. Based on simulations for a period of five years with an annual river flow regime based on the Malewa river in Kenya, the model showed that P retention was lower than N retention, leading to reduction of the N:P ratio in the water and an N-limited environment. Absence of surface water during the annual dry season caused a reduction of biomass through nutrient limitation. Harvesting increased N retention from 7% to over 40%, and P retention from 4% to 40%. Sensitivity analysis was done with a Monte Carlo procedure, drawing values for 28 model parameters drawn from rectangular probability distributions and evaluating the results for ten output variables related to biomass, nutrient concentrations and retention. The sensitivity analysis revealed that assimilation, mortality, decay, re-translocation, nutrient inflow and soil porosity were the most influential factors. Papyrus Simulator is suitable for studying nutrient retention and harvesting in wetlands, and contributes to quantification of ecosystem services and sustainable wetland management.

The concluding *Chapter 6* synthesizes the results from the preceding chapters and discusses these against the background of economic and agricultural development pressure in Africa. Until now, wetlands have been largely absent from earth system models, and models like Papyrus Simulator can contribute to a better assessment of the impact of wetlands on water quality and carbon storage. The model can also be used to evaluate the use of riparian buffer zones and constructed wetlands to mitigate the impact of agriculture on water quality. It has the potential to be coupled to spatially explicit regional or global hydrological models to quantify wetland ecosystem services on larger scales. The quantification of N and P retention can contribute to improving the analysis of trade-offs between provisioning and regulating ecosystem services of wetlands and to a more balanced economic valuation of wetland benefits, which currently often overemphasizes provisioning services and neglects the loss of regulating services when wetlands are converted to other uses like agriculture. In this way, the model can contribute to improving wetland policy (including the role of wetlands in achieving the Sustainable Development Goals of the United Nations) and wise use (as advocated by the Ramsar Convention on Wetlands), and support a more evidence-based adaptive management approach for African wetlands. Priority areas for further research to develop Papyrus Simulator include the belowground biomass and peat development in papyrus wetlands, which is relevant to the role of papyrus in carbon storage and climate change mitigation; greenhouse gas exchanges in papyrus, including the role of papyrus in denitrification; a more detailed description of the effects of climate and altitude on wetland processes; and the role of water limitation on papyrus growth. Options for further work with the model include the use of data from remote sensing and from aerial photography with drones for calibration and validation.

SAMENVATTING

Cyperus papyrus L. is een snelgroeiende zeggesoort die tot een lengte van 5 meter kan groeien en dominant is in de vegetatie van wetlands in een gebied dat zich uitstrekt van het Midden-Oosten, via Oost- en Centraal-Afrika tot in Zuidelijk Afrika. Papyrusmoerassen leveren al sinds mensenheugenis belangrijke ecosysteemdiensten voor miljoenen mensen: ze produceren water, vis en ander voedsel, medicijnen, en materialen voor allerlei toepassingen zoals de bouw van huizen en de productie van gebruiksvoorwerpen. Papyrusmoerassen zijn ook belangrijk door hun regulerende en culturele ecosysteemdiensten, en door hun biodiversiteit. In het landschap vormen ze buffers voor water en nutriënten, bieden ze habitat voor vissen, vogels en andere fauna, en vormen ze al eeuwen een integraal onderdeel van de levenswijze en cultuur van Afrikaanse moerasbewoners. Wetlands in Afrika staan onder druk van een toenemende bevolking en economische ontwikkeling, als gevolg waarvan het totale wetlandareaal afneemt. Een hogere voedselproductie zal gepaard moeten gaan met de bescherming van wetlands en hun ecologische integriteit, zeker nu de groeiende bevolking in Afrika de vraag naar voedsel doet toenemen. Ook eutrofiëring is in toenemende mate een probleem in deze regio, en papyrusmoerassen staan erom bekend dat ze de uitstroom van stikstof (N) en fosfaat (P) naar benedenstroomse gebieden kunnen verlagen. Voor het kwantificeren van de rol van moerassen in de retentie van N en P is meer kennis van de processen die aan retentie bijdragen nodig. De belangrijkste doelstelling van dit proefschrift is de ontwikkeling van een dynamisch simulatiemodel voor de kringlopen en retentie van nutriënten in papyrusmoerassen. Het uiteindelijk doel is om met een beter begrip van de processen van N en P retentie de effecten van verschillende beheersstrategieën (met name veranderingen in waterregime en het oogsten van papyrus) beter te kunnen voorspellen. *Hoofdstuk 1* verschaft achtergrondinformatie over papyrus-moerassen in Afrika en de ecosysteemdiensten die ze leveren, en geeft een overzicht van de overige hoofdstukken in dit proefschrift en hun doelstellingen.

In *Hoofdstuk 2* worden de invloed van periodieke overstroming en menselijke activiteiten op de retentie van N en P in moerassen met *Cyperus papyrus* onderzocht, met nadruk op de bovengrondse biomassa. Het onderzoek vond plaats in twee locaties in Oost-Afrika die zowel periodiek als permanent overstromen: Nyando Wetland in Kenia, dat wordt verstoord door landbouwactiviteiten en het oogsten van moerasvegetatie; en Mara Wetland in Tanzania, waar de verstoring door menselijke activiteiten kleiner is. De groei van individuele papyrusstengels en hun dichtheid werd gevolgd gedurende een periode van drie maanden door middel van een systematisch opgezet veldexperiment met kwadranten van 1 m^2 in zowel de periodiek als permanent overstroomde zones. De maximale groei van papyrusstengels werd goed beschreven (regressies voor de stengellengte met R^2-waarden tussen 0.70 en 0.99) met een logistisch model, waarbij stengels in Nyando sneller groeiden dan in Mara (r-waarden van respectievelijk 0.081-0.097 d^{-1} en 0.071-0.072 d^{-1}) maar korter bleven. De maximale stengellengte was hoger in permanent overstroomde (413 cm in Nyando, 484 cm in Mara) dan in periodiek overstroomde zones (287 en 464 cm, respectievelijk). De totale bovengrondse biomassa was hoger in Mara (4.4-7.5 kg m^{-2}) dan in Nyando (1.4-2.5 kg m^{-2}). De opgeslagen hoeveelheden N en P waren hoger in Mara dan in Nyando. Op basis van deze resultaten kon geconcludeerd worden dat papyrusplanten in verstoorde locaties kenmerken vertonen van r-geselecteerde soorten met een snelle groei als korte-termijn respons op een verstoring maar lagere biomassa en nutriëntenopslag op de langere termijn. In minder verstoorde locaties vertoont papyrus K-geselecteerde groeikenmerken met een aanvankelijk lagere groei maar uiteindelijk een hogere biomassa en nutriëntenopslag. Deze resultaten kunnen bijdragen aan het optimaliseren van de retentie van nutriënten in zowel natuurlijke als kunstmatige moerassystemen, bijvoorbeeld voor het bepalen van optimale oogststrategieën voor de planten.

In *Hoofdstuk 3* wordt, als eerste stap in de ontwikkeling van een model dat de retentie van nutrienten en het effect van het oogsten van *Cyperus papyrus* kan kwantificeren, een overzicht van bestaande moerasmodellen gepresenteerd. Dit hoofdstuk verkent de countouren van een model met als belangrijkste oogmerk om N en P retentie in papyrusmoerassen te bepalen, en hoe de onderliggende processen worden beïnvloed door zowel anthropogene als natuurlijke factoren. Een overzicht van de wetenschappelijke literatuur over wetlandmodellen resulteerde in 75 publicaties die werden onderverdeeld in vier categorieën: hydrologische modellen (17), biogeochemische modellen (20), vegetatiemodellen (28) en geïntegreerde modellen (10). Omdat de focus van het beoogde model ligt op de water-, nutriënten- en plantgerelateerde processen op wetlandschaal werd het overzicht vervolgens toegespitst op de publicaties over de biogeochemische en vegetatie-modellen, waarvan er tien werden geselecteerd die processen met N of P bevatten. Op basis van deze tien modellen werden de vereiste inputs voor een papyrusmodel geïdentificeerd, zoals de watertoevoer, regenval, evapotranspiratie, zonnestraling, nutrienten (N en P), groeikarakteristieken van de biomassa, oogsten van biomassa, en porositeit van de bodem. De mogelijke outputs van een model werden ook op een rijtje gezet: watervolume, waterdiepte, bovengrondse en ondergrondse biomassa, retentie van N en P, N- en P-concentraties in de afwatering en in het kwelwater. Vervolgens werd een conceptueel model gepresenteerd voor 1 m^2 papyrusmoeras met de genoemde inputs en outputs, aangedreven door een eenvoudige hydrologische component. De kenmerken van papyrus en de rol van het rhizoom in de opslag van nutriënten vereisen dat in het model een onderscheid wordt gemaakt tussen boven- en ondergrondse biomassa, en dat de interactie tussen N en P als potentieel limiterende nutriënten onderdeel is van het model. Tenslotte werd geconcludeerd dat een biomassamodel het meest geschikt is voor papyrus, gedeeltelijk omdat beperkte gegevens de ontwikkeling van een model voor individuele stengelgroei in de weg zouden staan.

In *Hoofdstuk 4* wordt de eerste versie van zo'n dynamisch simulatiemodel voor de N kringloop in natuurlijke niet-drijvende papyrusmoerassen in Oost-Afrika gepresenteerd. Het model bestaat uit sub-modellen voor de permanent en periodiek overstroomde zones en was gebaseerd op gegevens over een papyrusmoeras in Naivasha, Kenia. In beide zones werden de water-, N- en koolstofstromen berekend op basis van de hydrologische processen zoals het rivierdebiet, de waterdiepte in het meer, neerslag, verdamping; en ecologische processen zoals fotosynthese, N opname, mineralisatie, en nitrificatie. Literatuurgegevens afkomstig van het uitgebreide onderzoek dat sinds 1970 aan de papyrusmoerassen van het Naivasha-meer is uitgevoerd, aangevuld met gegevens uit publicaties over andere papyrusmoerassen, werden gebruikt voor de parameterisatie en calibratie van het model. Het model bestond uit 35 toestandsvariabelen en 143 snelheidsvariabelen en werd geïmplementeerd met behulp van de Stella software voor dynamische simulatie. Na een vergelijking met literatuurwaarden bleek dat het model realistische waarden berekende voor de concentraties opgeloste N en voor de biomassa van papyrus. Dagelijks oogsten tot een hoeveelheid van 84 g droge stof m^{-2} d^{-1} (in de periodiek overstroomde zone) en 60 g m^{-2} d^{-1} (in de permanent overstroomde zone) verlaagden de biomassa en verhoogden de N retentie (uitgedrukt als ($N_{instroom}$-$N_{uitstroom}$)/$N_{instroom}$ * 100%) tot 38% (in de periodiek overstroomde zone) en 50% (in de permanent overstroomde zone). Een nog hogere oogstintensiteit leidde tot de ineenstorting van de bovengrondse biomassa. De papyrusbiomassa herstelde daarentegen volledig van een jaarlijkse oogst van 100% van de biomassa. Het model toonde aan dat de herstelgroei van papyrus na oogsten in de permanent overstroomde zone N-gelimiteerd was omdat niet genoeg N beschikbaar was voor plantengroei door uitstroom naar het meer.

In *Hoofdstuk 5* wordt het model (dat nu 'Payprus Simulator' heet) verder ontwikkeld en wordt een gevoeligheidsanalyse uitgevoerd. Ook wordt een module voor fosfor (P) toegevoegd, en wordt het model meer algemeen toepasbaar gemaakt door het samenvoegen van de periodiek en permanent overstroomde zones (die op de situatie bij het Naivashameer waren gebaseerd) tot één generieke zone die permanent of periodiek overstroomd kan zijn afhankelijk van de rivierafvoer, het waterniveau van het meer en de neerslag. De belangrijkste outputvariabelen van het model (bovengrondse en ondergrondse biomasssa, N en P concentraties in biomassa en water, en netto biomassaproductie) werden vergeleken met gegevens uit een overzicht van 22 veldstudies met papyrusmoerassen. Simulaties voor een periode van vijf jaar met een jaarlijks rivierafvoerregime gebaseerd op de Malewa rivier in Kenia toonden aan dat de retentie van P lager was dan die van N, waardoor de N:P verhouding in het water daalde en een toestand van N-limitering ontstond. De afwezigheid van oppervlaktewater gedurende het jaarlijks terugkerende droge seizoen veroorzaakte een daling in biomassa door nutriëntenlimitering. Oogsten van vegetatie verhoogde de N retentie van 7% tot boven 40%, en P retentie van 4% tot 40%. Gevoeligheidsanalyse werd uitgevoerd met een Monte Carlo procedure, waarbij waarden voor 28 modelparameters werden getrokken uit uniforme kansverdelingen en de resultaten beoordeeld aan de hand van tien outputvariabelen zoals biomassa, nutriëntenconcentraties en retentie. De gevoelig-heidsanalyse liet zien dat assimilatie, mortaliteit, afbraak, retranslocatie, nutriëntentoevoer en bodemporositeit de meest invloedrijke parameters waren. Papyrus Simulator is geschikt voor het bestuderen van nutriëntenretentie en de effecten van het oogsten van vegetatie, en kan een bijdrage leveren aan het kwantificeren van ecosysteemdiensten en het duurzaam beheer van wetlands.

Het afsluitende *Hoofdstuk 6* bevat een synthese van de resultaten van de voorgaande hoofdstukken en plaatst die tegen de achtergrond van de noodzaak voor de ontwikkeling van economieën en landbouw in Afrika. Tot nu toe ontbreken wetlands goeddeels in aardsysteemmodellen, en modellen zoals de Papyrus Simulator kunnen een bijdrage leveren aan betere schattingen van de invloed die wetlands hebben op de waterkwaliteit en de opslag van koolstof. Het model kan ook gebruikt worden voor het beoordelen van bufferstroken en helofytenfilters wanneer die gebruikt worden om de impact van landbouw op de waterkwaliteit te verminderen. Het model kan gekoppeld worden aan ruimtelijk gedefinieërde regionale of globale hydrologische modellen om ecosysteemdiensten te kwantificeren op grotere schaal. Het kwantificeren van N en P retentie kan bijdragen aan het verbeteren van de analyse van zogenaamde 'trade-offs' tussen productiediensten en regulerende ecosysteemdiensten in wetlands, en aan een betere economische waardering van die diensten. Momenteel krijgen daarbij de productiediensten vaak veel aandacht terwijl het verlies van de regulerende ecosysteemdiensten bij het omzetten van wetlands in andere gebruiksvormen, zoals landbouw, wordt genegeerd. Op deze manier kan het model bijdragen aan verbetering van beleid rondom wetlands (inclusief de rol van wetlands bij het realiseren van de Sustainable Development Goals van de Verenigde Naties) en aan een duurzaam gebruik van wetlands ('wise use', zoals dat gepropageerd wordt door de Ramsar Conventie), en een meer op kennis gebaseerde adaptieve beheersaanpak voor Afrikaanse wetlands bevorderen. Prioriteiten voor de verdere ontwikkeling van de Papyrus Simulator liggen op het gebied van de ondergrondse biomassa en veenlagen in papyrusmoerassen, hetgeen van direct belang is voor de rol die papyrus speelt in de opslag van koolstof en het beheersen van klimaatverandering; de productie van broeikasgassen in papyrusmoerassen, incusief de rol van papyrus in denitrificatie; een meer gedetailleerde beschrijving van de effecten van klimaat en hoogteligging op de processen in wetlands; en het effect van waterlimitatie op de groei van papyrus. Toekomstig werk met het model zou gebruik kunnen maken van data uit remote sensing en van luchtfotografie met behulp van drones om het model te calibreren en valideren.

CONTENTS

Contents

CONTENTS

Contents

1

INTRODUCTION

1.1 THE IMPORTANCE OF WETLANDS

Wetland ecosystems are important for human well-being because of the benefits they provide to humans. These benefits are also known as ecosystem services and are a result of (but also depend on) the functioning of these ecosystems: their regulating functions in global water and nutrient cycles, and their high productivity and biodiversity (MEA 2005; TEEB 2010; Diaz et al. 2015; Darwall et al. 2018). Another important reason for conserving wetlands and their ecosystem services is climate change. Wetlands can increase the resilience to climate change by storing carbon, contributing to cooling of the climate in the long term, and by storing water and providing a buffer against both droughts and flooding (McInnes 2018; Moomaw et al. 2018). As a result, the estimated economic value of wetlands is proportionally much higher than their modest share of global land-use (Russi et al. 2013). Because of their ecosystem services and climate change benefits, wetlands are important for achieving the Sustainable Development Goals (SDGs) of the United Nations (Ramsar convention 2018b).

Globally, wetlands are under pressure and disappear at an alarming rate due to land use change, conversion to farmland or settlements, water abstraction and diversion, and overexploitation of wetland products (Davidson 2014; Ramsar Convention on Wetlands 2018b). It is estimated that about two-thirds of the world's natural wetlands have disappeared since 1900 (Davidson 2014; Davidson et al. 2018), and this loss is still continuing (van Asselen et al. 2013; Dixon et al. 2016). These pressures not only lead to wetland loss, but also influence the potential of remaining wetlands for delivering ecosystem services. Therefore, to sustainably benefit from wetlands, it is now widely accepted that a balanced use of different ecosystem services is needed, and that trade-offs, e.g. between food production and water quality regulation, need to be recognized and incorporated into wetland management policy and planning (De Groot et al. 2018). Several international conventions and organizations are now promoting this concept of sustainable use or 'wise use', such as the Ramsar Convention on Wetlands, the UN Framework Convention on Climate Change, the UN Convention to Combat Desertification, and the UN Convention on Biological Diversity, the Food and Agriculture Organization of the UN, and UN-Water. Between 13 and 18% of the world's wetlands are listed on the Ramsar Convention list of wetlands of international importance (Davidson and Finlayson 2018). Despite these efforts to conserve and protect wetlands, effectively implementing sustainable management regimes is still a challenge (Finlayson 2012).

1.2 PAPYRUS WETLANDS IN AFRICA

In Sub-Saharan Africa, where population and food demands are growing, wetlands dominated by the sedge *Cyperus papyrus* L. are increasingly overexploited and converted to satisfy these demands (Kipkemboi and van Dam 2018). Before the 20th century, papyrus wetlands were abundant and widely spread in central and eastern Africa, ranging from the Okavango delta to the Nile valley and from Benin to Madagascar, but also in the Middle East and Europe (Kipkemboi and van Dam 2018). At present, papyrus is still widespread in the same regions, but less abundant, especially in the downstream part of the river Nile (van Dam et al. 2014; Carballeira and Souto 2018), and increasingly reported to be under pressure of conversion and exploitation elsewhere (Owino and Ryan 2007; Namaalwa et al. 2013; Maclean et al. 2014). Papyrus plants (Figure 1.1) consist of a rhizome that supports 5-6 culms with umbels (Thompson et al.1979) that can reach lengths of 5-6 m (Gaudet 1977b). They exhibit clonal and sexual reproduction (Gaudet 1977a). Papyrus is a C4 photosynthesis species (Jones 1988) and is known for its rapid growth and productivity: after cutting the aboveground biomass, it can grow back fully in a period of about six to nine months (Muthuri et al. 1989; Terer et al., 2012b). These wetlands can be found mainly in the floodplains of rivers and lakes, and dominate large areas when undisturbed, for example 1,100 km^2 around the Upemba basin in Congo, 2,500 km^2 in the Okavango in Botswana (Thompson and Hamilton 1983), and 3,900 km^2 in the Sudd wetland in South Sudan (Howell et al. 1988). There are also many smaller papyrus wetlands of importance, e.g. in the valley bottoms of Rwanda and Uganda, some of which are listed as Ramsar sites (Denny 1984; 1985; Kipkemboi and van Dam 2018). The total area of papyrus wetlands is not known, due to a lack of monitoring information (Adam et al. 2014).

The earliest evidence of papyrus use dates back to about 4,600 years in ancient Egypt where papyrus was used to build ships and to produce paper (Ch. Munch 1861). Today papyrus wetlands are still of crucial importance for livelihoods of millions of people (Morrison et al. 2012). Through direct use of papyrus (Figure 1.2) for building houses, crafts, fish traps, fuel (Ojoyi 2006; Osumba et al. 2010), brooms and to produce alcohol (observation, November 16, 2010), but also by using fertile soils in the dry season to produce crops and land for livestock grazing (Rongoei et al. 2013). Other benefits include provisioning of drinking water, sand and clay for brick making (Kibwage et al. 2008) and providing habitat for fish as well as acting as a nursery for larger fish species (Gichuki et al. 2001; Kiwango et al. 2013). The wetlands also regulate water quality and quantity, store carbon and regulate the local climate, play a role in local religious and spiritual activities (Kibwage et al. 2008) and are appreciated for tourism and aesthetics (Maclean et al. 2011; Ajwang' Ondiek et al. 2016).

Figure 1.1 Pictures of Cyperus papyrus: overview of Mara Wetland (left top); root and rhizome of mature plant (right top); young culm (left bottom); floating papyrus island (right bottom)

Pressures on wetlands are a global problem, not least in Africa where population growth, the need for food security, the suitability of wetlands for food production and weak implementation of wetland conservation policies create an enormous pressure on wetlands, while at the same time people depend directly on the many ecosystem services of these ecosystems (Rebelo et al. 2010; Mitchell 2013). The ever increasing demand for land to produce food leads to conversion of wetlands into cropland and grazing areas (van Asselen et al. 2013; Beuel et al., 2016), yet at the same time wetlands are crucial for storing and providing water and nutrients to the agricultural sector, provide a crucial role in sustaining (commercial) fisheries and regulate the local climate on which especially the poor depend (Silvius et al. 2000). While in theory the delivery of provisioning services (e.g. through harvesting of fish or through wetland agriculture) can be balanced with regulating services (de Groot et al. 2010), in practice the alterations to the hydrological, ecological and biogeochemical functions due to agriculture are often permanent and destructive (Kansiime et al. 2007). Another cause for imbalance is that agriculture as a provisioning service is more tangible, and often prioritized in decision making over a regulating service such as water quality regulation, the value of which decision-makers may not always appreciate (Kipkemboi and van Dam et al. 2018). As a result, papyrus wetland area has been declining, in line with the dramatic global reduction of wetland area (Davidson et al. 2018). For example in the Nile basin, area reduction was estimated at approximately 7% in the period from the mid-eighties to the start of the 21st

Figure 1.2 Pictures of ecosystem services: scientists looking for data (left top); papyrus fish trap (right top); seasonal agriculture (left bottom); harvested papyrus culms (right bottom)

century (Maclean 2014). Despite these negative trends, papyrus wetlands offer potential for sustainable use and a balanced utilization of their ecosystem services because they are highly productive and recover quickly from disturbances (van Dam et al., 2014). While being threatened, they are still abundant and crucial to sustain the growing population of the region.

1.3 TRADE-OFFS IN PAPYRUS WETLAND ECOSYSTEM SERVICES

Research in support of trade-off analysis of ecosystem services in papyrus wetlands, and a better understanding of the impact of converting wetlands to other uses, on their water quality regulation function (especially nitrogen and phosphorus retention) are important prerequisites for implementing sustainable wetland management. Nitrogen (N) and phosphorus (P) runoff to the numerous freshwater lakes of East Africa (e.g. Lake Naivasha and Lake Victoria) leads to eutrophication, prolific growth of water hyacinth and algal blooms (Nyenje et al. 2010; Olokotum et al. 2020). This has a direct impact on people as fish catch declines, drinking water quality deteriorates and tourists stay away. Degradation and loss of papyrus wetlands in the riparian zones may worsen eutrophication, as these fringing wetlands often function as buffer zones (Fisher and Acreman 2004). While intermediate levels of livelihoods activities like seasonal agriculture and cattle grazing in wetlands may be sustainable, high levels of exploitation can lead to the degradation and destruction of these wetlands (Rongoei et al. 2014; MacLean et al. 2014). This does not only negatively impact downstream waterbodies, but also makes the area less suitable for crop production and cattle grazing because water storage and local rainfall may be reduced. This makes papyrus wetlands a highly relevant social-ecological system for research, especially as there is a large scope for strengthening the implementation of sustainable wetland management strategies in Africa (Langan et al. 2018). As many wetlands in Africa are still intact but under increasing pressure, implementing sustainable wetland management now would be a better strategy than losing wetlands to economic development first and then having to restore wetland ecosystem services later on, as they are for example essential in support of rain-fed agriculture (Gordon et al. 2010; Rockström and Falkenmark 2015).

It is difficult to analyse the trade-off between food production and nutrient retention in wetlands without understanding the underlying processes involved and how they are impacted, either positively or negatively. The processes, components and structure of papyrus wetlands that are responsible for retaining or removing the different forms of N and P are affected by human activities as well as by natural drivers like hydrology and climate. So far, studies on the ecosystem functions of papyrus wetlands and how they are affected by both natural and anthropogenic drivers have been limited. The main processes involved in retaining N and P are biological or physico-chemical in nature. N and P are retained by trapping particulate matter and sediments (Boar and Harper 2002) and by storing both living and dead organic matter (Gaudet 1977a, Gaudet and Muthuri 1981; Boar et al. 1999). N can also be permanently removed through denitrification (van Dam et al. 2007, Gettel et al. 2012), while P can be adsorbed to sediment particles (Kelderman et al. 2007). The efficiency of papyrus wetlands in retaining or removing N and P depends on the hydrology and how a natural wetland is connected to rivers, lakes and groundwater (van Dam et al. 2014). Besides natural papyrus ecosystems, work on constructed wetlands with papyrus has also been a considerable source of knowledge about ecological processes and nutrient retention in papyrus wetlands (Abira et al. 2005; Mburu et al. 2015; Sepúlveda et al. 2020).

1.4 MODELLING OF NUTRIENT RETENTION IN PAPYRUS WETLANDS

Decision-making about the impact of provisioning services on regulating services can be supported by explanatory models that describe retention of N and P in wetland ecosystems (Maltby 2009; Carpenter et al. 2009). Wetland models that describe the hydrology, growth of emergent vegetation and the processes related to N and P cycling can be used as a starting point for developing a more specialized papyrus wetland model. Many freshwater ecosystem models were developed to increase understanding and to aid and improve management of wetlands (Janssen et al. 2015). Depending on the questions the model tries to answer, wetland models focus on either hydrology (e.g. Fan and Miguez-Macho 2011; Hughes et al.

2014), biogeochemistry (e.g. Wang and Mitch 2000; Melton et al. 2013) or vegetation (e.g. Benjankar et al. 2011; Wang et al. 2013). Some models combine two or even all of these aspects (e.g. Zhang et al. 2002; Xu et al. 2016). Another category of models includes socio-economic aspects and integrates these to analyse ecosystem responses to anthropogenic drivers (e.g. Feng et al. 2011; van Dam et al. 2013). For the research described in this thesis, the focus will be on the components (soil, vegetation) and processes (hydrology, biochemistry) of papyrus wetlands and how they respond to changes in human activities, such as harvesting of papyrus.

For the development of a model for trade-off analysis in papyrus wetlands, the main components and processes related to retention of N and P need to be described. Existing lake and wetland models were reviewed that describe the same or similar components and processes (Mooij et al., 2010; 2019). Van der Peijl and Verhoeven (1999) developed a model describing carbon, N and P dynamics in a temperate river marginal wetland to simulate the effect of human influences on these dynamics. A similar approach was used in PCLake to model N and P cycling in shallow lakes with marsh areas in the Netherlands (Sollie et al. 2008). Asaeda (2008) presents an overview of organ specific growth models of aquatic plants like *Phragmites* spp. and *Typha* spp. and focuses on the translocation between below- and aboveground systems. This is especially important when the aboveground biomass is harvested and the belowground parts remain in the wetland. Van Dam et al. (2007) published a simulation model of N retention in a floating papyrus wetland with data from Kirinya wetland in Jinja, Uganda. This model is suitable as a foundation of a new papyrus model, but lacks processes related to P and does not include papyrus that is rooted in soil.

1.5 RESEARCH OBJECTIVES AND CHAPTER INTRODUCTIONS

The overall objective is to develop a dynamic simulation model for papyrus wetlands that can support the analysis of trade-offs between a reduction in the area covered by *Cyperus papyrus* for food production or biomass harvesting (provisioning ecosystem services) and N and P retention (regulating ecosystem services). The specific objectives are described for each chapter:

Chapter 2 presents the results of field experiments related to growth of papyrus and N and P uptake. The experiments were carried out in permanently and seasonally flooded zones in Nyando wetland, Kenya and Mara wetland, Tanzania both fringing Lake Victoria. The chapter presents the growth and uptake rates of N and P of *Cyperus papyrus* at different growth stages of the individual culms. It also discusses how much N and P is stored in living papyrus biomass, and it looks into the differences between the sites and their hydrology. Finally it shows how the empirical data can be used to calculate the impact of harvesting on uptake rates and retention.

Chapter 3 reviews the literature on existing ecosystem models describing processes and components related to N and P cycling in wetlands, and how these are influenced by both anthropogenic and natural drivers. The chapter explains the pros and cons of the different models and explains the choices that were made in relation to the development of a model for papyrus wetlands.

Chapter 4 describes the first stage of the development of this simulation model for rooted papyrus. The chapter explains how the model was constructed for aboveground and belowground papyrus biomass in permanently and seasonally flooded wetland zones. Then it shows the parametrization and calibration of the model based on data from Lake Naivasha, Kenya. Finally, it calculates biomass growth, N concentration in the water, and N retention for the two hydrological zones and for different vegetation harvesting regimes.

In Chapter 5 the model (now called 'Papyrus Simulator') is further developed into a model that simulates N and P retention together. The process to incorporate P, to improve the hydrological section and to make it more generally applicable was described. A global sensitivity analysis is presented as well as a model validation. Finally the model was used to simulate retention in the different zones as well as under different harvesting scenarios and to show the impact of on N to P ratios which are important in relation to the effects of eutrophication.

Finally in chapter 6 an overview of the main findings is presented. The chapter draws main conclusion in relation the overall objective and gives recommendations, both scientific as well as in relation to the societal relevance and application of the work.

1.6 SIGNIFICANCE OF THE STUDY

Provisioning services often bring the most direct local benefits and are also the most easy to quantify and value. They are often prioritized over benefits that are more difficult to measure (Brauman et al. 2007; Brander et al. 2013), but their economic value is generally much lower than the value of the regulating services (TEEB 2010; de Groot et al. 2012). Therefore it is important to understand how to quantify other benefits (e.g., water quality regulation), and especially their reduction in relation to (over)exploitation of provisioning services (Silvius 2000, Raudsepp-Hearne et al. 2010, Mouchet et al. 2014). A better understanding of these trade-offs can help to maximize benefits in a sustainable way (Brauman et al. 2007; Carpenter et al. 2009; Raudsepp-Hearne et al. 2010). Haberl et al. (2006) state this understanding as being essential for integrating the socioeconomic dimension into long term ecological research, which is crucial for conservation.

Since the introduction of the Millennium Ecosystem Assessment report in 2003 and its 'wetlands and water synthesis' in 2005, the call for practical application of the ecosystem services framework has increased further (Carpenter et al. 2009; Guerry et al. 2015). This thesis tries to answer this call with the construction of a model that allows quantification of ecosystem services in East African papyrus wetlands. The research starts with generating new knowledge based on field experiments, and then combines the field data with a simulation model to increase understanding of N and P cycling in papyrus systems. This is important for quantifying the impact of harvesting on N and P retention and concentrations in natural papyrus wetlands, but could also support the design and operation (e.g. harvesting regimes, wastewater loads) of constructed papyrus wetlands for wastewater treatment.

Achieving the sustainable development goals (SDGs) requires wetland conservation, sustainable use and even restoration. According to the Ramsar convention, wetlands contribute to the achievement of all 17 SDGs (Ramsar 2018). This research and the development of the Papyrus Simulator contributes to a more sustainable use and conservation of papyrus wetlands and through that supports the achievement of the SDGs in general. More specifically, the focus on balanced use of these systems and a better understanding of the underlying processes will facilitate sustainable agriculture, livestock and fisheries, thus contributing to SDG2 (Zero hunger). Moreover, water quality regulation contributes to SDG 6 (clean water and sanitation); and the maintenance of habitat and storage of carbon contribute to SDG13 (climate action), SDG14 (life below water) and SDG15 (life on land) (Janse et al. 2019).

2

THE EFFECT OF SEASONAL FLOODING AND LIVELIHOOD ACTIVITIES ON RETENTION OF NITROGEN AND PHOSPHORUS IN *CYPERUS PAPYRUS* WETLANDS, THE ROLE OF ABOVEGROUND BIOMASS[1]

ABSTRACT

With growing demand for food production in Africa, protecting wetlands and combining increased agricultural production with conservation of the ecological integrity of wetlands is urgent. The role of aboveground biomass of papyrus (*Cyperus papyrus*) in the storage and retention of nitrogen (N) and phosphorus (P) was studied in two wetland sites in East Africa under seasonally and permanently flooded conditions. Nyando wetland (Kenya) was under anthropogenic disturbance from agriculture and vegetation harvesting, whereas Mara wetland (Tanzania) was less disturbed. Maximum papyrus culm growth was described well by a logistic model (regressions for culm length with R^2 from 0.70 to 0.99), with culms growing faster but not taller in Nyando than in Mara. Maximum culm length was greater in permanently than in seasonally flooded zones. Total aboveground biomass was higher in Mara than in Nyando. The amounts of N and P stored were higher in Mara than in Nyando. In disturbed sites, papyrus plants show characteristics of r-selected species leading to faster growth but lower biomass and nutrient storage. These findings help to optimize management of nutrient retention in natural and constructed wetlands.

Keywords: nutrient regulation, regulating ecosystem services, trade-offs, constructed wetlands, agriculture, water quality

---

[1] Published as:

Hes EMA, Yatoi R, Laisser SL, Feyissa AK, Irvine K, Kipkemboi J, van Dam AA (2021) The effect of seasonal flooding and livelihood activities on retention of nitrogen and phosphorus in *Cyperus papyrus* wetlands, the role of aboveground biomass. Hydrobiologia https://doi.org/10.1007/s10750-021-04629-3

2.1 INTRODUCTION

By 2050, it is expected that the African population will grow to 2.5 billion (UN, 2015). This will increase the demand for food, and the need to increase agricultural production. As the productivity of African agricultural systems is low compared with other world regions, increased production is often achieved by areal extension (van Asselen et al., 2013; Schoumans et al., 2015; OECD/FAO, 2016) at the expense of natural ecosystems like forests, natural grasslands and wetlands (UNCCD, 2017; Ramsar Convention on Wetlands, 2018; FAO, 2019). This has an impact on biodiversity, and soil and water quality (Butchart et al., 2010; UNCCD, 2017). It also poses risks for human well-being, as food production and a good living environment depend on the regulating and provisioning services of healthy, well-functioning ecosystems (MEA, 2005; IPBES, 2019). In Africa, wetlands contribute to water quality regulation and flood protection (Silvius et al., 2000; Schuyt, 2005; Verhoeven & Setter, 2010), and high wetland carbon storage may help with climate change mitigation (IPCC, 2014). Moreover, wetlands can increase the resilience of poor rural communities against climate change effects like floods and droughts and have relatively high economic value and biodiversity (Russi et al., 2013; Darwall et al., 2018; Tickner et al., 2020). There is thus a trade-off between agricultural production and the loss of ecosystem services. As agricultural development in Africa progresses, protecting wetlands and combining increased agricultural production with conservation of the ecological integrity and function of wetlands is an urgent need (Jayne et al., 2019).

Utilisation of wetlands to meet human needs has implications on ecological functions. A key ecological function affected when wetlands are converted or degraded is nutrient and sediment retention (Johnston, 1991). Wetlands often serve as buffer zones in the landscape, and their degradation, in combination with fertilizer application and soil erosion in the catchment, can increase nutrient and sediment runoff into rivers and lakes (Hecky et al., 2010). In the Lake Victoria region in East Africa, harvesting of papyrus (*Cyperus papyrus*) vegetation and seasonal agriculture in papyrus wetlands are widespread (Kipkemboi & van Dam, 2018). This has led to increased sediment and nutrient loads, contributing to ecological degradation of Lake Victoria with algal blooms and excessive growth of the water hyacinth (*Eichhornia crassipes*) with serious economic consequences (Kiwango & Wolanski, 2008; Hecky et al., 2010; Olokotum et al., 2020).

Nutrient and sediment retention by papyrus wetlands involves a number of biological and physical processes. Nitrogen (N) and phosphorus (P) removal from surface water include temporary storage in above- and belowground biomass and adsorption to soil particles, and longer term storage in deeper peat layers (Gaudet, 1977; Kansiime et al., 2007). Permanent removal of N by the wetland ecosystem occurs only through denitrification (Galloway et al., 2004; Pina-Ocha & Álvarez-Cobelas, 2006). Both natural and anthropogenic disturbance can interfere with these processes, and change the nutrient retention characteristics of the wetlands. For example, natural disturbances include flooding and drying cycles, and grazing by herbivores. Anthropogenic disturbances include livelihood activities like seasonal or permanent conversion to agriculture and vegetation harvesting. To illustrate this, drainage will lower the moisture content of the soil, which has consequences for adsorption properties and denitrification. It will also affect the growth of the papyrus plants, which changes the flow and storage of nutrients in above- and belowground biomass. Different patterns of harvesting and removal of aboveground biomass have an impact on nutrient retention (Terer et al., 2012; Chapter 5). Some areas are cleared of papyrus vegetation entirely for seasonal agriculture in the dry season, after which the vegetation grows back when the wetland floods again in the rainy season (van Dam et al., 2013). In other areas harvesting is more selective because different uses (e.g. fibre for making chairs, roof thatch, mat making and other household crafts) require different grades of papyrus culms. When the belowground biomass of the plants is left intact, the fast regrowth of papyrus can restore the aboveground biomass in about 6-9 months (Terer et al., 2012), demonstrating a high resilience of this system. As pressure from harvesting and agriculture increases, it is important to understand the recovery of the aboveground biomass and associated N and P accumulation under different utilization regimes.

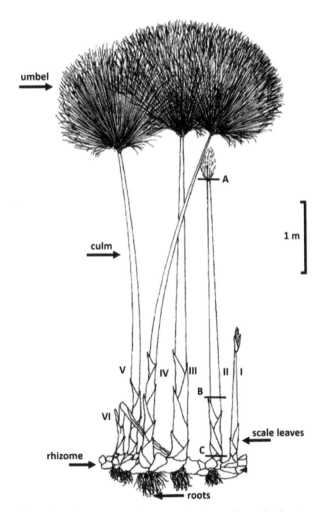

Figure 2.1 Papyrus with development stages: I=closed umbel; II=opening umbel; III=fully opened umbel; IV=flowering; V=mature; VI senescing and points for measurements: A=base of umbel; B=top of scale leaves; C= base of culm. Figure adapted by permission from the Licensor: Springer Nature, Economic Botany (Muthuri & Kinyamario, 1989)

Research quantifying the role of aboveground biomass in nutrient balances is limited to modelling (van Dam et al., 2007; Chapter 5) and measurements on constructed wetlands (Kengne et al., 2008). To understand how much N and P are stored aboveground, the biomass density (in g dry weight [DW] m^{-2}) and the N and P content of biomass (in % DW) need to be quantified. Reported productivity of aboveground papyrus biomass is 5-37 g DW m^{-2} d^{-1}, aboveground biomass 1384-6045 g DW m^{-2}, and N and P content 0.65-1.75 % DW and 0.024-0.13 % DW, respectively (reviewed in Chapter 5). Immature papyrus shoots have a higher N and P content than mature plants (Gaudet, 1977; Rongoei & Kariuki, 2019), and this influences the amounts of N and P present in aboveground biomass. How these characteristics change over time depends on how the growth rates of shoots respond to the prevailing environmental conditions (e.g., flooding levels, water quality, harvesting, grazing). Papyrus culms can grow very fast (up to 4-5 m within about 6 months) and typically exhibit six vegetative growth stages (Figure 2.1). While the impact of environmental conditions on biomass and density of papyrus has been studied (Gaudet, 1975; Muraza et al., 2013; Opio et al., 2017; Geremew et al., 2018), there is no data on the effect of disturbances on N and P content of culms of the different growth stages.

Growth of papyrus biomass (culm and umbel) was described and simulated by a logistic model, with estimated values for maximum biomass and instantaneous growth rates based on literature values (van Dam et al., 2007; Opio et al. 2014). The amounts of N and P stored in aboveground biomass can be estimated by combining the logistic growth model with the N and P content at different growth stages. The overall aim of this study was to quantify the relationship between aboveground papyrus biomass and retention of N and P to better manage the balance between use and conservation of natural systems and improve performance of constructed wetlands for wastewater treatment. The three specific objectives were to: (1) describe growth of aboveground biomass with a logistic model and estimate the model parameters in sites with different disturbance conditions (flooding, intensity of human use); (2) compare standing biomass, growth of aboveground biomass and uptake of N and P under different disturbance conditions; and (3) assess the role of living aboveground biomass in retention of N and P under different conditions and identify implications for sustainable use. Our hypothesis is that papyrus culm growth follows a logistic model, N and P content decreases while the culms develop from young to mature, and that culm development and density is influenced by natural conditions and anthropogenic pressures. Based on two growth strategy theories (r/K selection and CSR triangle) we expect disturbed and drier areas will have faster growth, lower biomass and less fully matured plants compared with undisturbed wet areas (Pianka, 1970; Grime, 1977). As a result, undisturbed areas would have higher biomass, but lower N and P content. Therefore, there may be more N and P stored in mildly disturbed areas with lower biomass and higher N and P content. To explore this, we conducted a field experiment in natural papyrus wetlands in Kenya and Tanzania.

2.2 METHODS

2.2.1 DESCRIPTION OF THE FIELD SITES

The field experiments were carried out during the period November 2010 – January 2011 in Nyando wetland in Kenya and Mara wetland in Tanzania (Figure 2.2). Nyando wetland (0°11'-0°19' S to 34°47'-34°57' E) borders Lake Victoria at Winam Gulf at an altitude of 1140 m and is adjacent to Kenya's third biggest city, Kisumu within the district Kisumu East, with a population of 475,000 at the time of this study (KNBS, 2010). The total wetland area was estimated at 30-50 km^2 (Khisa et al., 2013) and the intact part is largely dominated by papyrus. Mara wetland (1°27'-1°37' S to 33°55'-34°28' E) borders Lake Victoria at the mouth of the Mara River at an altitude of 1134 m and is a papyrus-dominated wetland of more than 350 km^2 (Bregoli et al., 2019). The population in and around the wetland was 56,000, divided over 20 villages (The United Republic of Tanzania, 2013). Just upstream of the wetland is a mining area (North Mara Gold Mine) and an associated town, Nyamongo. Both Nyando and Mara are similar papyrus dominated floodplain wetlands and surrounded by predominantly agricultural land used for crops and cattle grazing. While the overall anthropogenic pressure in Nyando leads to a reduction of wetland area (Khisa et al., 2013 and Rongoei et al., 2013), in the Mara the wetland is expanding (Bregoli et al., 2019). Comparing the state of the two field sites we observed that papyrus growth in general is better in Mara with less anthropogenic disturbance (harvesting and agriculture) and larger undisturbed permanently flooded areas.

In both sites, two experiments were carried out. The first experiment monitored the growth of aboveground biomass (culms). In the second experiment, biomass, culm density and nitrogen and phosphorus content of the plants were measured. Both experiments were done in permanently and in seasonally flooded zones dominated by papyrus. The permanently flooded zone was defined as saturated soils with standing water all year around and the seasonally flooded zone was not fully saturated all year round. Because of the different conditions in the two locations, there were slight differences in experimental set-up in both sites that are described below.

2.2.2 MONITORING OF ABOVEGROUND BIOMASS GROWTH

In Nyando wetland, two transects were cut through the papyrus dominated vegetation stands: one in Singida at Bwaja River bordering the agricultural zone, and another in Ogenya at the shore of Winam Gulf (Figure 2.2 C,E,F). In Mara wetland, three transects were cut perpendicular to the main channel of Mara River (Figure 2.2 D,G). Because of time and resource limitations, we could only monitor the first and the third transect (Mara 1 and Mara 3) for this experiment. All transects were made from the seasonally flooded zone to the permanently flooded zone, perpendicular to the open water (Nyando) or the river (Mara). In both zones of each transect, a 10 m path was cut perpendicular to the transect (Figure 2.2).

The development stages of the papyrus plants were assessed using the description by Muthuri & Kinyamario (1989). Along the transect paths, four (Nyando) or five (Mara) culms of each development stage were selected randomly and marked with a waterproof marker (Figure 2.2). The development of these individual culms was followed for eight weeks. At the start, and then every two weeks, the following measurements were taken: culm length (cl, in cm); culm length from the top of the scale leaf (clsl, in cm); culm width at the top of the scale leaf (width, in cm); culm girth at the top of the scale leaf (girth, in cm). Culm lengths were measured with a wooden stick and tape measure, culm width with a vernier calliper, and culm girth with a length of rope and a tape measure.

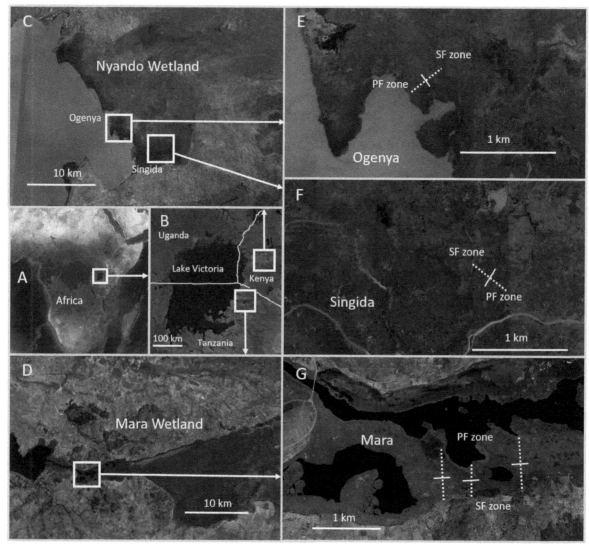

Figure 2.2 Locations of field sites and transects for sampling with: A=Africa; B=Lake Victoria; C=Nyando Wetland; D=Mara wetland; E=Ogenya sampling site; F=Singida sampling site; G=Mara sampling site. In EF and G: white dashed line=transect; solid line=transition between zones; SF-zone=seasonally flooded zone; PF-zone=permanently flooded zone (satellite images taken from Google Earth Pro at 17 June 2020)

To describe general conditions, the following water quality measurements were taken at the start of the experiment, after 4 weeks and after 8 weeks: electrical conductivity (EC, in µS/cm), temperature (in °C), pH, dissolved oxygen concentration (DO, in mg/L) (all with a model 3210 SET multimeter; Wissenschaftlich-Technische Werkstätten, Hamburg, Germany), ammonium-nitrogen (NH$_4$-N in mg/L), nitrate-nitrogen (NO$_3$-N in mg/L), soluble reactive phosphorus (PO$_4$-P in μg/L), total nitrogen (TN in mg/L) and total phosphorus (TP in mg/L). All N and P analysis followed standard methods (APHA, 1992). Nyando analysis were carried out in Njoro, Kenya at Egerton University. In the Mara site, only pH, DO, NH$_4$-N, NO$_3$-N and TN were measured, due to limitations in the Musoma, Tanzania laboratory.

2.2.3 BIOMASS AND CULM DENSITY

In Nyando, two quadrats of 1x5 m were selected in each zone (eight in total) along the same transects (Figure 2.2). In Mara, three 1x5 m quadrats were selected in each zone of transects 1 and 3, and two quadrats in both zones of transect 2 (16 in total). In all quadrats, the entire aboveground biomass was harvested by cutting the culms just above the rhizome. The culms were counted and classified according to their development stage. Culms were initially sun-dried, and then oven-dried at 80°C to determine their dry weight.

2.2.4 NITROGEN, PHOSPHORUS AND CARBON CONTENT IN ABOVE- AND BELOWGROUND PLANT ORGANS

In each of the Nyando 1x5 m quadrats, two 1x1 m quadrats (Figure 2.2) were harvested for both above- and belowground biomass. All culms were cut just above the rhizome and divided into development classes (Figure 2.1), counted and measured (lengths, width and girth as described above). After sun and oven drying, the combined dry weight of all culms of the same development stage was measured. Then, the plant material was cut in pieces of approximately 2 cm, mixed and one sample for each development stage for each zone was taken to determine nitrogen (N), phosphorus (P) and carbon (C) content. The rhizomes and roots were sun- and oven-dried to constant weight and dry weight was determined. After cutting into 2 cm pieces and mixing, a sub-sample was taken to determine N, P and C content. For Mara wetland the same procedure was followed with two 1x1 m quadrats in the seasonally and permanently flooded zones of transects 2 and 3 (Figure 2.2).

All samples were analysed at the laboratory of the Chemical Biological Soil Laboratory of Wageningen University, the Netherlands. After grinding the samples to 1 mm, N and P (mg/g DW) were determined by digestion of the samples with H$_2$SO$_4$/Se/salicyclic acid and H$_2$O$_2$ and segmented flow analysis (SFA).

2.2.5 DETERMINING LOGISTIC GROWTH MODELS

The growth pattern of the papyrus can be described using the logistic growth equation:

$$\frac{dL}{dt} = r * L \left(1 - \frac{L}{K}\right)$$
(Equation 1)

With L = culm length in cm), t = time (day), r = instantaneous growth rate (day^{-1}) and K = the maximum culm length (cm). Dividing the equation (Equation 1) by L gives:

$$\frac{dL/dt}{L} = -\frac{r}{K} * L + r$$
(Equation 2)

a linear relationship with the relative growth in length (dL/dt /L) as dependent variable, length (L) as the independent variable, intercept r and slope r/K. For each 2-week growth period of the culms, individual relative culm growth (L_2-L_1)/days and mean culm length $((L_1+L_2)/2)$ were calculated. For the culms showing maximum growth, model II linear regression lines were estimated between mean culm length (X) and relative growth rate (Y) (Figure 2.3), using the lmodel2 package in R. The parameters r and K of the logistic curve were calculated for each transect from the intercept and slope (Rongoei & Outa, 2016). The same method for determining the logistic growth models was used for width and girth.

2.2.6 DATA ANALYSIS

Confidence intervals (95%) for the parameters of the logistic growth curves (r, K) were estimated based on the standard error estimates for the regression coefficients (Sheather, 2009). Differences in mean culm length, biomass, culm density, and N and P content among growth stages were compared using one-way Type III analysis of variance (ANOVA). Within growth stages, differences in the same variables among transect locations (Ogenya, Singida, Mara) and inundation zones (seasonal or permanent) were compared using two-way Type III ANOVA. All variables were checked for normal distribution and log10-transformed when necessary (biomass, culm density). All statistical tests were performed using R version 4.0.2 (R Core Team, 2020).

2.3 RESULTS

All regression models were significant, but more significant for both culm length and culm length from top of scale leaf than for width and girth. R^2 values were highest for culm length between 0.89 and 0.99 and one of 0.70 (Table 2.1 and Figure 2.3), especially compared with culm length from top of scale leaf. The regression models based on culm length were therefore used to estimate the carrying capacity (K, maximum length) and the instantaneous growth rate (r) (Figure 2.3).

Table 2.1 Adjusted R^2 and significance (*** p<0.001; ** p<0.01; * p<0.05; · p<0.1) for regression models based on: culm length (cl), culm length from top of scale leave (clsl), width and girth; PF is permanently flooded and SF is seasonally flooded

Site	cl		clsl		width		girth	
	R^2	p	R^2	p	R^2	p	R^2	p
Ogenya PF (n=6-17)	0.99	***	0.91	***	0.90	***	0.79	*
Ogenya SF (n=5-10)	0.98	***	0.94	***	0.84	*	0.80	**
Singida PF (n=7-11)	0.92	***	0.80	***	0.87	**	0.85	***
Singida SF (n=5-13)	0.70	**	0.87	***	0.81	*	0.89	**
Mara PF (n=8-19)	0.91	***	0.60	***	0.99	***	0.91	***
Mara SF (n=13-16)	0.89	***	0.67	***	0.91	***	0.96	***

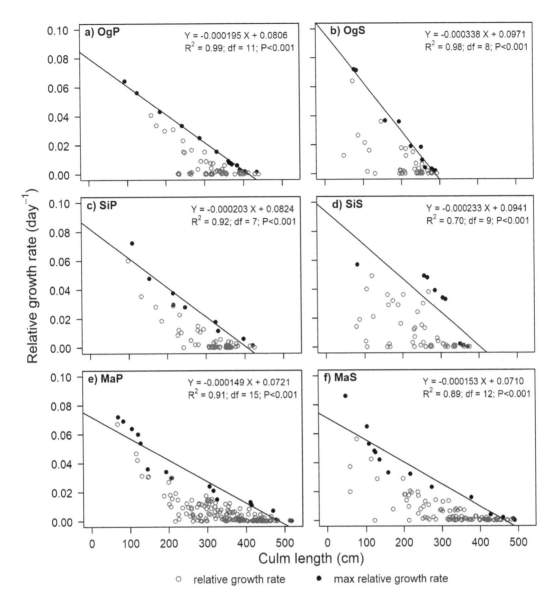

Figure 2.3 Relative growth rate (day-1) in relation to culm length of papyrus. Data points represent 2-week growth intervals for individual culms among three sites and two zones per site with OgP = Ogenya permanent flooded zone (n=77); OgS = Ogenya seasonally flooded zone (n= 54); SiP = Singida permanently flooded zone (n=70); SiS = Singida seasonally flooded zone (n=68); MaP = Mara permanently flooded zone (n=187) and MaS = Mara seasonally flooded zone (n=119). Solid circles = maximum relative growth rate and open circles = relative growth rate. Model II regression lines were calculated for the points with maximum growth. For details see text

The results of the regression models based on culm length showed the highest instantaneous growth rate (0.097 d^{-1}) in Ogenya seasonal flooded zone. Growth rates were all higher in Nyando (between 0.081 and 0.097 d^{-1}) than in Mara (0.071 d^{-1} in Mara permanently flooded zone and 0.072 d^{-1} in Mara seasonally flooded zone). In both Nyando sites, the growth rates were higher in the seasonally flooded zones (0.091 d^{-1} in Ogenya seasonally flooded zone and 0.081 d^{-1} in Ogenya permanently flooded zone ; 0.094 d^{-1} in Singida seasonally flooded zone and 0.082 d^{-1} in Singida permanently flooded zone) and in Mara there was no difference between the zones (Figure 2.4). The maximum culm length (K) was higher in the permanently flooded zones for each site, most notably in Ogenya (413 and 287 cm respectively). In Mara (484 and 464 cm respectively), the maximum culm length was higher than in Nyando (between 287 and 413 cm). Based on the confidence intervals, the growth difference between permanent and seasonal flooding was only significant (p<0.05) in Ogenya (Figure 2.4).

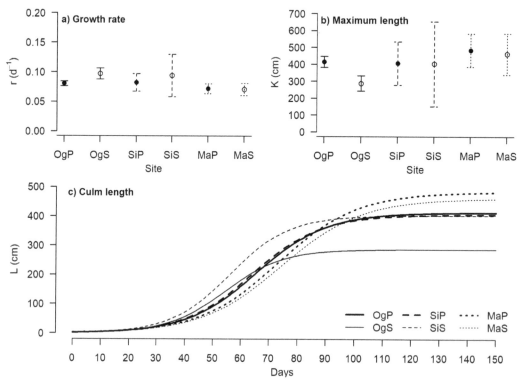

Figure 2.4 Estimated growth parameters of the logistic curve for papyrus in six transects. a) Instantaneous growth rate r; b) maximum length K; and c) culm length over time among three sites and two zones per site with OgP=Ogenya permanently flooded zone, OgS=Ogenya seasonally flooded zone, SiP=Singida permanently flooded zone, SiS=Singida seasonally flooded zone, MaP=Mara permanently flooded zone and MaS=Mara seasonally flooded zone. Error bars in a) and b) indicate 95% confidence intervals

Culm length increased significantly from growth stage I to growth stage IV (Figure 2.5). In both Singida and Ogenya, the average culm length was higher in the permanently flooded zone than in the seasonally flooded zone. In the most mature phase (V), the culms in Mara were significantly longer than in both Nyando sites (Figure 2.5a). In all locations, there was no significant difference between the amount of biomass classified as stage I, II, III or IV per m^2. However, the highest amount of biomass was in growth stage V, especially for Mara and within Mara in the permanently flooded zone (Figure 2.5b). Density for most stages was below 5 culms per m^2, with no difference between seasonally - or permanently flooded zones. The stage V (mature culms) counts were higher, especially in Mara ($\approx$ 15 culms m^{-2}, while stage IV (flowering) was low or absent in Mara and always present in both Nyando sites (Figure 2.5c). The content of N and P in the culms were highest during stage I and II and decreased with every consecutive stage, a pattern seen in all the sites or zones (Figure 2.6a & b).

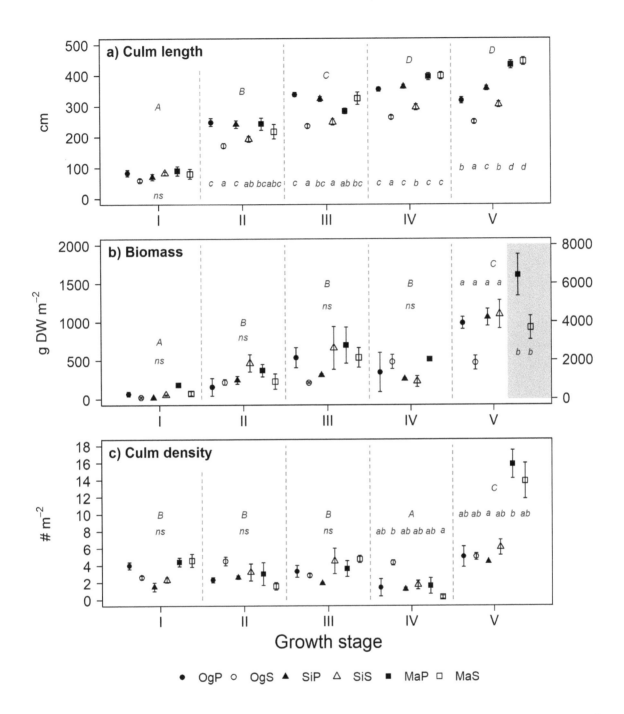

Figure 2.5 Field measurements of a) culm length; b) biomass and c) culm density per growth stage among three sites and two zones per site with OgP=Ogenya permanently flooded zone, OgS=Ogenya seasonally flooded zone, SiP=Singida permanently flooded zone, SiS=Singida seasonally flooded zone, MaP=Mara permanently flooded zone and MaS=Mara seasonally flooded zone. Note: in b) the two data points in the gray rectangle (MaP & MaS) were plotted on a different scale, see second y-axis. Error bars indicate standard error. Growth stages sharing the same capital letter were not significantly different (1-way Type III ANOVA, p < 0.05). Within growth stages, sites sharing the same lower case letter were not significantly different (2-way Type III ANOVA, p < 0.05)

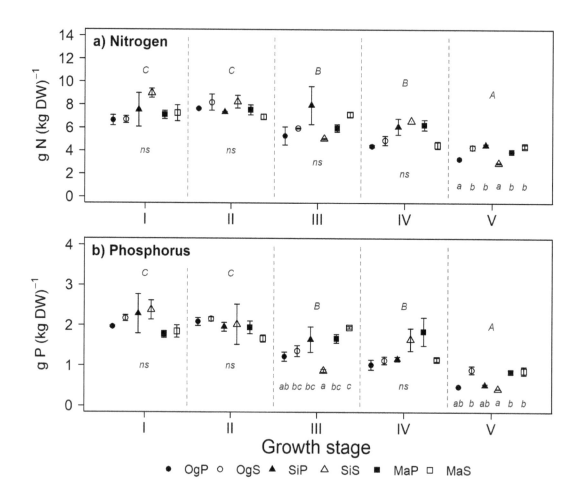

Figure 2.6 Nitrogen a) and phosphorus b) content of culm per growth stage among three sites and two zones per site with OgP=Ogenya permanently flooded zone, OgS=Ogenya seasonally flooded zone, SiP=Singida permanently flooded zone, SiS=Singida seasonally flooded zone, MaP=Mara permanently flooded zone and MaS=Mara seasonally flooded zone. Data points are means of composite culm samples (2 for Ogenya and Singida, 4 for Mara) taken from vegetation quadrats. Error bars indicate standard error. Growth stages sharing the same capital letter were not significantly different (1-way Type III ANOVA, p < 0.05). Within growth stages, sites sharing the same lower case letter were not significantly different (2-way Type III ANOVA, p < 0.05)

The culm density, amount of biomass and average length of all culms of growth stage I to V were highest in the permanently flooded zone of Mara wetland, followed by Mara seasonally flooded zone. In both Singida and Ogenya the culm density was higher in the seasonally flooded zone than the permanently flooded zone, while the average length was higher in the permanently flooded zone than the seasonally flooded zone (Figure 2.7).

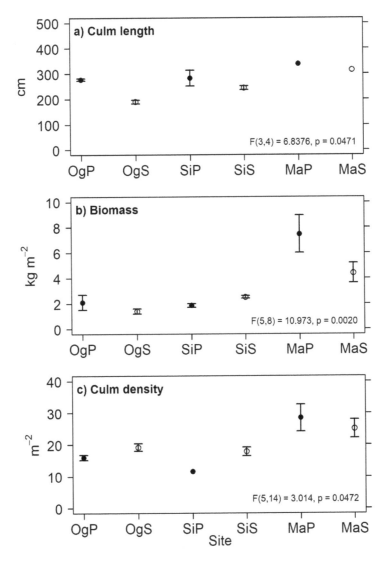

Figure 2.7 Comparison of average culm density, culm biomass and culm length among three sites and two zones per site with OgP=Ogenya permanently flooded zone, OgS=Ogenya seasonally flooded zone, SiP=Singida permanently flooded zone, SiS=Singida seasonally flooded zone, MaP=Mara permanently flooded zone and MaS=Mara seasonally flooded zone. Error bars indicate standard error, except for MaP and MaS in a) as length for all culm classes together in Mara was not measured, but calculated based on culm density per class and average length per class

As a result of the higher culm density and biomass, the amount of N and P per m^2 was also higher in Mara compared with the Nyando sites (Table 2.2). In Mara around two thirds of the N and P was stored in mature culms (growth stage V), while in Nyando this was more evenly spread over the growth stages II-V (Table 2.2).

Table 2.2 Amount and distribution of nitrogen and phosphorus in standing biomass [g/m2] in OgP=Ogenya permanently flooded zone, OgS=Ogenya seasonally flooded zone, SiP=Singida permanently flooded zone, SiS=Singida seasonally flooded zone, MaP=Mara permanently flooded zone and MaS seasonally flooded zone

	OgP	OgS	SiP	SiS	MaP	MaS
N I	0.5	0.1	0.1	0.5	1.3	0.5
N II	1.2	1.8	1.8	3.8	2.8	1.6
N III	2.8	1.3	2.5	3.4	4.2	3.8
N IV	1.5	2.4	1.6	1.5	3.2	0.0
N V	3.2	2.0	4.7	3.3	25.0	16.3
N I-V	**9.3**	**7.6**	**10.7**	**12.5**	**36.5**	**22.2**
P I	0.1	0.0	0.0	0.1	0.3	0.1
P II	0.3	0.5	0.5	0.9	0.7	0.4
P III	0.6	0.3	0.5	0.6	1.2	1.0
P IV	0.3	0.5	0.3	0.4	0.9	0.0
P V	0.5	0.4	0.5	0.4	5.1	3.3
P I-V	**1.9**	**1.8**	**1.9**	**2.5**	**8.3**	**4.9**

Table 2.3 Amount and nitrogen and phosphorus content of belowground biomass (BGB) in OgP=Ogenya permanently flooded zone, OgS=Ogenya seasonally flooded zone, SiP=Singida permanently flooded zone, SiS=Singida seasonally flooded zone, MaP=Mara permanently flooded zone and MaS=Mara seasonally flooded zone. Error is standard error

	OgP (n=2)	OgS (n=2)	SiP (n=2)	SiS (n=2)	MaP (n=3-8)	MaS (n=3-7)
BGB [g/m^2]	964 ±75	1564 ±501	1886 ±98	1452 ±31	26638 ±5180	10144 ±2193
N BGB [mg/g DW]	7.9 ±0.9	5.1 ±0.3	4.8 ±0.6	5.7 ±0.2	7.7 ±2.3	4.5 ±1.0
P BGB [mg/g DW]	1.3 ±0.1	1.7 ±0.2	0.9 ±0.0	1.3 ±0.0	2.4 ±0.3	1.4 ±0.3
detritus [g/m^2]	32 ±0.5	47 ±9.2	31 ±10.4	33 ±3.0	477 ±128	360 ±80
N detritus [mg/g DW]	4.6 ±0.1	5.4 ±0.5	5.8 ±1.1	6.1 ±0.6	9.8 ±1.5	8.7 ±1.0
P detritus [mg/g DW]	0.5 ±0.0	0.7± 0.1	0.8 ±0.2	0.8 ±0.0	1.0 ±0.2	0.8 ±0.2

The belowground (root and rhizome) and detritus biomass were highest in the permanently flooded zone in Mara and then Mara's seasonally flooded zone. Values for Ogenya and Singida were considerably lower in both zones and similar with each other (Table 2.3). Content of N and P in the belowground biomass were similar to the values in the aboveground biomass (Figure 2.6) and Table 2.3), with no significant differences among the sites or zones. Nitrogen values in the detritus were similar to content in living biomass, while the values for phosphorus were lower (Table 2.3).

Oxygen concentrations in Ogenya and Singida seasonally flooded zones were around 2 mg/L, which was higher than in the permanently flooded zones and in both zones in Mara (all below 1 mg/L). The pH was around 6 in all sites and zones, except for Singida permanently flooded zone (7.5 ± 2.1). The water temperature was around 21°C in both Mara sites, around 22°C in the Nyando permanently flooded sites and around 24.5 °C in the Nyando seasonally flooded sites (Table 2.4). Ammonium concentration was highest in Singida (0.31 ± 0.09 mg/L) and nitrate was higher in Mara than in both Nyando sites. Total N was around 1 mg/L in all sites.

Table 2.4 Selected water quality parameters in OgP=Ogenya permanently flooded zone, OgS=Ogenya seasonally flooded zone, SiP=Singida permanently flooded zone, SiS=Singida seasonally flooded zone, MaP=Mara permanently flooded zone and MaS=Mara seasonally flooded zone. Error is standard error.

WQ	OgP (n=4-22)	OgS (n=4-22)	SiP (n=4-22)	SiS (n=4-22)	MaP (n=7-21)	MaS (n=13-21)
DO (mg/L)	0.30 ±0.05	2.52 ±0.37	0.72 ±0.06	1.89 ±0.22	0.29 ±0.02	0.27 ±0.02
NH_4-N (mg/L)	0.18 ±0.07		0.31 ±0.09		0.14 ±0.01	0.15 ±0.01
NO_3-N (mg/L)	0.002 ±0.001		0.002 ±0.001		0.06 ±0.00	0.09 ±0.01
TN (mg/L)	1.04 ±0.18		1.01 ± 0.15		1.00 ±0.17	0.83 ±0.03
pH	6.1 ±0.43	6.1 ±0.13	7.5 ±1.05	6.4 ±0.19	6.0 ±0.02	6.0 ±0.03
Temp (°C)	22.4 ±0.59	24.6 ±0.31	21.6 ±0.64	24.6 ±0.55	21.0 ±0.15	21.0 ±0.13

2.4 DISCUSSION AND CONCLUSION

Papyrus culms grew faster, but not as tall, in the Nyando sites (Singida and Ogenya), compared with the Mara site. The maximum culm length (K) in all sites was greater in the permanently flooded zones than in the seasonally flooded zones. The growth rate (r) in the seasonally flooded zones in the Nyando sites were higher compared with the permanently flooded zones in Nyando, and with both zones in Mara. Biomass was higher in Mara compared with Nyando, but culm density was not significantly different. The regression models used to estimate r and K were based on the relation between maximum growth and length, and describe a growth pattern for each site. Culms growing at a slower rate were not used for estimating the maximum growth, as these are likely limited by nutrient availability or light availability due to self-shading (Jones, 1988; Saunders et al., 2014). The models were highly significant, with high coefficients of determination (R^2 >0.89), confirming that the growth pattern of papyrus culms can be adequately described using a logistic growth model. The only exception was the seasonally flooded zone in Singida (R^2=0.70), which had fewer measurements of longer culms which may have reduced the coefficient of determination. The logistic growth pattern is well known from other studies on papyrus (Kansiime et al., 2003; Opio et al., 2014), but also for other emergent macrophytes such as *Phragmites australis* (Zemlin et al., 2000; Clevering et al., 2001) and *Typha domingensis* (Lorenzen et al., 2001; Lagerwall et al., 2012).

Culm growth was measured with non-destructive methods as proxies for biomass increase. A standard method, like the diameter at breast height (DBH) for trees, is not available for emergent macrophytes like papyrus. Therefore different methods were tried. Culm length from the scale leaf, width at the scale leaf, and girth at scale leaf were measured, however culm length measured from the base of the culm at the rhizome (Figure 2.1) gave the best results. Possibly the growth rate of the scale leaf is different from that of the whole culm as competitive plants are known to change allometry under stress (Grime, 1977), which could influence the consistency of the three other methods. Besides, for girth and width small measuring mistakes result in larger errors, which could be reduced by the use of high quality digital Vernier calipers. Another improvement in growth parameter (r and K) estimates could be realized by more frequent (weekly, instead of bi-weekly) measurements (Deegan et al., 2007), and by including more culms of growth stages I and II in the experiment to have more data points at lower culm lengths. There may be seasonal differences in growth rates or resource allocation (Pianka, 1970; Grime, 1977). The experiments ran in between the so called short and long rainy seasons, although because of variability in the rainfall patterns in Nyando and Mara the seasons are not always very distinct (Gabrielsson et al., 2013). According to Opio et al. (2017), productivity of papyrus was not affected much by seasonal variations.

The results of the growth analysis indicate that zones with different disturbance levels show differences in papyrus growth strategy. With higher pressure from livelihood activities such as harvesting, grazing and cropping (Nyando) and with seasonal flooding, culms grow faster but shorter. Where disturbance is less (Mara) and with permanent flooding, culms grow slower but taller. This is consistent with theories that characterize species based on their growth strategies, and growth and recovery rates (Soissons et al., 2019). One of those theories is r/K selection, where 'K' refers to carrying capacity and 'r' to the intrinsic rate of natural increase. Rapid development, high growth rates, early reproduction, and small body size are traits of r-selected species. K-selected species are characterised by slower development, great competitiveness, delayed reproduction, and larger size (Pianka, 1970). A second theory is the CSR triangle, where species are classified as Competitive (C), Stress Tolerant (S), Ruderal (R), or as a combination. S is the equivalent of K-selected and R of r-selected, and C species combine traits of the others (Grime, 1974; 1977). The vegetation response in Nyando (growing faster, but less tall) is characteristic of C-R species which under favourable conditions grow fast (clonal growth), dominate large areas and make large and rapid changes in their allometry and allocation in response to stress (Weiner, 2004) for example by increased seed production. When harvested, burned or grazed (in the seasonally flooded zones, and more so in Nyando than in Mara) they re-grow quickly, but shorter, following the classic strategy of an r-selected species. Under light competition and when close to carrying capacity (the permanently flooded zones, especially in Mara) the papyrus showed more traits of K-selected or R species: slow, tall, and strong (Pianka, 1970).

When looking at the culm densities per growth stage, a similar growth strategy emerges. Under higher pressure from livelihood activities (Nyando), there is more seed production (stage IV), characteristic of R type plants. However, clonal reproduction was observed in both Nyando and Mara, and allowed for fast spreading and dominance, characteristics of a competitive (C) strategy. The relatively high density of mature culms (stage V) in Mara and the lower number of flowering plants (stage IV) indicate a combined C-R strategy. This was confirmed by findings of experiments under longer periods of stress, e.g. droughts and high sediment loads, where sexual reproduction and seedling recruitment become a more prevalent growth strategy for papyrus plants (Terer et al., 2014; Geremew et al., 2018).

Mara wetland was a more productive environment than Nyando, as evidenced by the higher biomass per m^2. The total biomass in Mara (4.4-7.5 kg m^{-2}) was in the upper range, and in Nyando (1.4-2.5 kg m^{-2}) in the lower range of values for papyrus (1.4-8.7 kg m^{-2}; summarized in Chapter 5). This confirms that conditions were less affected by livelihood activities in Mara and that papyrus follows a C strategy under good conditions, outcompeting other species with fast clonal growth (Geremew et al., 2018). Nyando, however had a higher aboveground biomass production rate compared with Mara. The biomass was far below the carrying capacity, with growth, therefore, closer to the exponential part of the logistic growth curve than Mara. The observed differences in total aerial biomass between the more affected Nyando site and the more pristine conditions in Mara were in line with the findings in Terer et al. (2012) indicating that frequent harvesting impacts re-growth of papyrus and leads to a lower overall biomass.

Nitrogen and phosphorus content of the shoots at all sites was highest in stages I and II and decreased with every growth stage (III to V), with no significant differences among the sites or zones. The pattern of higher N and P content in earlier development of the shoots confirms earlier findings (Gaudet, 1977; Muturi & Jones, 1997), and can be explained by active translocation from the rhizome to support fast culm development (Rejmánaková, 2005; Asaeda et al., 2008). The amount of N found in aboveground biomass was at the lower end, and P at the higher end of the range reported in the literature (0.65-1.75 %DW for N; 0.024-0.13 %DW for P; summarized in Chapter 5). Lake Victoria has received increasing P loads since the 1960s (Hecky et al., 2010), which may have increased P uptake and storage of papyrus in Lake Victoria's riparian zones. The N and P content in the belowground biomass was comparable with the values in the aboveground biomass, with no significant differences among the sites or zones. In detritus, the P content was lower than in the living biomass, but N content was similar to that in living biomass. P release from dead culms was reported to be faster than N release, both in papyrus (Gaudet, 1977) and in other macrophytes (Chimney et al., 2006). This is likely from faster leaching of P, especially in the tropics under humid conditions, while N decomposition depends more on microbial processes (Manzoni et al., 2010).

As a result of the higher aboveground biomass, the N and P amounts per m^2 were higher in Mara than in Nyando. Most of the N and P in Mara was located in the mature (stage V) culms. In the Nyando sites there was a more equal distribution over the different growth stages. The amount of N stored in both Nyando sites was low (7-12 g N m^{-2}). Even the values found in Mara (22-36 g N m^{-2}) were lower than reported from other studies (48-104 g N m^{-2}; Gaudet, 1977; Boar et al., 1999; Boar, 2006). For Nyando, this can be explained by the low biomass. For Mara, the exclusion of senescing culms (Stage VI) from our analysis may explain the lower N values. The lower amount of P in the Nyando sites (1.8-2.5 g P m^{-2}) compared with reported values (5.4-5.7 g P m^{-2}; Gaudet, 1977; Boar, 2006) are likely because of relatively low biomass. The relatively high values in Mara P (4.9-8.3 g P m^{-2}) may be a result of increased P loading to the lake, as mentioned above. The study by Boar (2006) was carried out at Lake Naivasha, and the experiments described by Gaudet (1977) were carried out in the early 1970s, when the P loading to Lake Victoria was lower.

Natural and anthropogenic disturbance influenced N and P retention by aboveground biomass in our field sites. Based on different conditions, papyrus adopted different growth strategies and this influenced nutrient retention. Retention was higher in permanently inundated zones than seasonally inundated zones, and higher in sites with low disturbance (Mara) than in sites with livelihood activities (Nyando). Under ideal conditions, a K-selected or Competitive growth strategy leads to large mature culms, leaving less space for young culms and storing nutrients in the aboveground biomass. Under stressful conditions there is a shift to more r-selected growth, with young shoots growing fast, but attaining a lower height (Pianka, 1970), retranslocation of N and P to the rhizome for storage, and morphological flexibility with smaller shoots and denser rhizomes (Geremew et al., 2018). Under severe or long term stress, seed production can increase, ensuring survival after e.g. a long absence of water (Terer et al., 2014). The flexibility to switch between growth strategies makes papyrus wetlands resilient to environmental and anthropogenic pressure, and amenable to management measures for conservation and restoration (Morrison et al., 2012).

The results provide input for a wider application of modelling N and P retention (Chapter 5). Incorporating the effects of disturbance and the variation of N and P content of the culms can improve the estimation of regulating services like nutrient retention and water purification, and improve our understanding of trade-offs related to provisioning services such as food production (agriculture and fisheries). Quantitative models can contribute to the land-sharing/land-sparing debate (Kremen, 2015). Currently, agricultural development in eastern Africa follows a 'sharing' pattern, in which papyrus wetlands are used for extensive dry-season agriculture, grazing and harvesting of culms from which they recover in the wet season when access is limited. A 'sparing' approach would conserve papyrus areas with no or very limited anthropogenic pressures, but intensify crop production in designated agricultural zones. Sparing could have a positive impact on other ecosystems services including food production, biodiversity and general ecosystem integrity (Kiwango & Wolanski, 2008; Pacini et al., 2018), whereas sharing, with frequent harvesting, may be beneficial for optimizing or balancing a limited number of ecosystem services such as wetland based agriculture and N and P retention (Chapter 5). Modelling can support this debate by quantifying N and P impacts on water quality in different land-use scenarios, specifically through quantifying the impact of harvesting and other livelihood activities on retention and long term effects on regeneration of aboveground biomass (Terer et al., 2012).

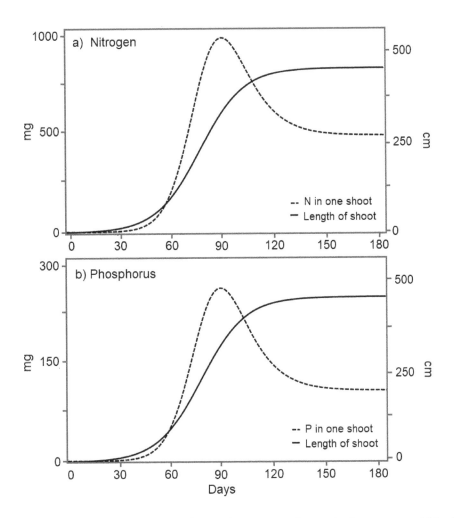

Figure 2.8 Nutrient content and length of one shoot over time: a) nitrogen; and b) phosphorus. Length was plotted with a logistic equation with average r (0.081 d^{-1}) and K (414.2 cm) of all sites. N and P were plotted by multiplying length with an estimated 2nd polynomial equation describing the relationship between average length of one shoot and N or P per shoot per growth stage of all sites, with N and P in one shoot as the dependent variable. Equation for N: y=-7*10^{-5} x^2 + 0.0318x in which y=N [mg N/cm] and x=culm length [cm], R^2 =0.88; and for P: y=-2*10^{-5} x^2 + 0.0089x in which y=P [mg P/cm] and x=culm length [cm], R^2 =0.67. Graphs were drawn with Stella Professional Version 1.7.1, ISEE systems and calculations made with MS Excel Professional Plus 2013

Another application for these results can be found in constructed wetlands for wastewater treatment (Kengne et al., 2008; Perbangkhem & Polprasert 2010). Our results can improve management of removal efficiency. A simulation of N and P in culms (Figure 2.8) showed that the maximum amount of N and P per culm was reached before the culm reached maximum length and after approximately 90 days. This was a result of increasing biomass (growth) and decreasing N and P concentrations (Figure 2.6), probably caused by active retranslocation of N and P even before maximum length was reached (Snyder & Rejmánková 2015). If harvesting is done when culms reach the maximum amount of N and P stored (and not when they are fully grown), N and P removal through harvesting can increase by almost 50%. Environmental conditions, higher nutrient loading, controlled flow and residence time are different in constructed wetlands and likely lead to differences in growth and N and P content (Chale, 1987; Kengne et al., 2008; Perbangkhem & Polprasert, 2010). The results of this study can help optimize N and P removal for a constructed wetland system with known growth rates and N and P content per growth stage, by increased frequency of harvesting and higher amount of N and P removed per harvest.

In conclusion, our findings can help to understand and quantify the impacts of livelihood activities and inundation on retention of N and P in natural wetlands, as well as optimize management for removal of N and P from wastewater by harvesting aboveground papyrus biomass in constructed wetlands. We recommend further research and model development on the role of biomass and accumulation of organic matter in N and P retention, and to incorporate our findings in local contexts to support management of both natural and constructed wetlands.

3

Review of Wetland Models for the Development of a Model to Quantify Nutrient Retention and the Impact of Harvesting in *Cyperus papyrus* Wetlands[2]

Abstract

African papyrus wetlands are important in provisioning crucial ecosystem services like food and clean water. However, they are also under pressure from human activities and economic development. As a result, the total area of these systems is declining. Eutrophication is of growing concern in the region and papyrus wetlands are known to reduce nitrogen (N) and phosphorus (P) loads to downstream water bodies. An increased ability to quantify the role of wetlands in N and P retention and a better understanding of the role of the different processes contributing to N and P retention is needed. This chapter explores the outlines of a model to assess N and P retention in papyrus wetlands, and how this is affected by both natural and anthropogenic drivers. A search of scientific literature revealed 75 publications on wetland models. These publications were divided in four categories: hydrological models (17), biogeochemical models (20), vegetation models (28) and integrated models (10). From the publications on biogeochemical and vegetation wetland models, ten models were selected that all included N or P processes. These ten models were used to identify required inputs (water inflow; rainfall; evapotranspiration; solar radiation; N and P inputs; biomass growth characteristics; biomass harvesting; soil porosity) and outputs (water volume; inundation levels; biomass amounts above- and belowground; N and P retention; N and P outputs). A conceptual model was proposed for 1 m^2 of papyrus wetlands with these in- and outputs and a relatively simple hydrological component as forcing factor.

Keywords: wetland modelling, nutrient retention, biogeochemical models, vegetation models

---

[2] This chapter is based on part of the work in Janse JH, van Dam AA, Hes EMA, de Klein JJM, Finlayson CM, Janssen ABG, van Wijk D, Mooij WM, Verhoeven JTA (2019) Towards a global model for wetlands ecosystem services. Current Opinion in Environmental Sustainability 36:11-19

3.1 INTRODUCTION

Wetland ecosystem services are crucial for human wellbeing. In Africa, many people depend directly on water and food provisioning of papyrus wetlands (Chapter 1). One of the many benefits of wetlands is the regulation of water quality, more specifically the role of wetland processes in the cycling of: phosphorus (P), nitrogen (N) and several other nutrients. Despite their benefits, wetlands globally are under pressure from human activities and economic development, and papyrus wetlands are no exception (Chapter 1). A good understanding of the ecosystem processes required to provide ecosystem services is needed, to enable sustainable management, policy development, and the determination of trade-offs between economic development and wetland conservation. Models of wetland processes and of the impact of natural and anthropogenic pressures on the role of wetlands in the N and P cycle can help quantify the retention of these nutrients, bring focus to field research, and support trade-offs and decision-making. This chapter explores the benefits and challenges of modelling nutrient retention (with emphasis on N and P) in wetlands. It reviews existing wetland models and suggests a modelling approach for N and P retention in African papyrus wetlands.

Temporal and spatial dynamics and complexity present challenges for managing the benefits (ecosystem services) and the adverse impacts (disservices) of wetlands to human well-being. Wetlands change continuously as a result of natural variation (seasonal and long-term variability, changing climate) and through human activities (land use, livelihood activities). These continuous changes alter the functioning of wetland ecosystems. For example, wetlands can change from being a net sink of carbon (C), through C sequestration, to being a source (through carbon dioxide or methane emission) and vice versa (Mitsch et al. 2013). Another example is the retention of N and P, with N and P that is stored in soil organic matter (or peat layers) under water-logged conditions and released when the wetland is drained or becomes drier due to changes in local climate (Moomaw et al. 2018).

Ecosystem modelling can be used to deal with both the complexity and the dynamics of aquatic ecosystems. This was demonstrated for natural systems (Mooij et al. 2010; Janssen et al. 2015), rice crops (Li et al. 2015), and aquaculture ponds (Reid et al. 2020). These reviews show how a variety of models can improve our understanding of complexity, as well as complement empirical research. Global models describing earth surface, vegetation and integrated assessment have informed policy and decision making on biodiversity, climate and food security (Erb et al. 2017; Bonan and Doney 2018; Pongratz et al. 2018).

The role of wetlands in these global models has been largely ignored. With the exception of methane emissions (Melton et al. 2013) and C sequestration (Nakayama 2017), these global models mostly exclude wetland-specific hydrological, biogeochemical and vegetation processes. While existing regional wetland models do not provide global coverage, several wetland modelling studies on specific types or regions have been helpful for decision-making.

These gaps in modelling need to be addressed to allow assessment of local, regional and global contributions of wetlands in fighting the global water and food crisis.

Papyrus wetlands are part of larger riverine and lacustrine catchments (Chapter 1). Development of a papyrus wetland model could contribute to evaluating trade-offs between provisioning ecosystem services (e.g. agriculture) and regulating ecosystem services (e.g. water quality regulation). The development of a papyrus model raises questions such as: What are the system boundaries? Which hydrological-, biogeochemical- and vegetation processes of the N and P cycle need to be included? What data is available? Which type of modelling (dynamic or static; spatial or non-spatial) should be used? What kind of environmental factors are important (e.g. rainfall, temperature, oxygen levels), and how should they be incorporated in a model?

The answers to these questions can be found in two places: the scientific literature on papyrus wetlands (which was summarized in Chapter 1); and existing wetland ecosystem models. The overall goal of this chapter is to explore the outlines of a model to assess N and P retention in papyrus wetlands, and how this is affected by both natural and anthropogenic drivers. This leads to the following specific objectives : (1) provide an overview of existing wetland models in the literature; (2) describe how the processes underlying N and P retention are modelled; and (3) identify an approach for modelling retention of N and P in papyrus wetlands.

3.2 METHODS

A literature survey was done using Scopus and Google Scholar, using the keywords 'wetland', 'model', 'hydrological', 'biogeochemical' and 'vegetation'. The resulting publications were sub-divided into four main categories, according to the emphasis in model objectives or structure: (a) hydrological models, focusing on water flows; (b) biogeochemical models, focusing on nutrient processes; (c) vegetation models, focusing on plant growth and productivity; and (d) integrated models, without a distinct focus on one of the first three categories or with a strong social science or economic component. Sometimes there were overlaps between and among categories with models containing aspects of several categories, but generally it was possible to assign models to one category. The models were assessed and compared based on the following criteria: model type, wetland type, input variables, output variables, scale, and application.

Based on the literature survey, a subset of models was selected from the biogeochemical and vegetation models (categories b and c). For this subset, a more detailed analysis was done of the methods used for quantifying biogeochemical and ecological processes, and N and P retention. Based on this more detailed analysis, conclusions were drawn about options for modelling N and P retention in papyrus wetlands.

3.3 RESULTS AND DISCUSSION

The literature survey resulted in a set of approximately 75 publications from the period 1985 - 2017. The publications presented a variety of approaches, from purely empirical models to explanatory dynamic simulation models. Several models were designed for specific wetland types or regions or had an otherwise restricted scope. Often, models were validated against an (independent) set of field data, but few were applied outside their development domain.

The hydrological models (Category a, 17 publications; Appendix 3.1) generally operated on a catchment scale or considered wetland hydrology within a catchment, which is required to adequately model the runoff and groundwater flows and crucial for understanding the hydrological functions of wetlands within the landscape, such as infiltration of precipitation and surface flow, water storage, groundwater recharge and discharge. Water table depth is a key variable determining the processes in wetlands and in wetland soils particularly (Fan and Miguez-Macho, 2011), as it determines the type of vegetation, the availability of oxygen, and the redox potential for processes like decomposition of organic matter, denitrification and nitrification. Some of these hydrological models focus on sediment processes and erosion (Day et al. 1999; Evrard et al. 2010; Schindewolf and Schmidt 2012). A further review of the hydrological processes included in the models is beyond the scope of this review.

Depending on their objectives, the integrated wetland models (Category d, 10 publications; Appendix 3.2) incorporated hydrological, biogeochemical and/or vegetation processes at varying levels of detail, or extended the analysis to socio-economic aspects or ecosystem services delivery. In terms of nutrient retention processes they did not add new insights for the research question of this chapter, and therefore were also excluded from further in-depth analysis here.

Table 3.1 lists the biogeochemical models identified in the survey (Category b, 20 publications), including descriptions of the model- and wetland type, and the input and output variables. Most models were process-based models. Two papers were more a synthesis of global data for modelling, rather than describing a model per se. There was a variety of wetland types, from inland to coastal wetlands (including mangroves and coastal freshwater wetlands), a few models for constructed wetlands, and some models for peat wetlands. Most models focused on the scale of one wetland, but there were some with a global spatial scale. The input variables for these models included factors influencing growth and decay of vegetation (e.g. temperature; irradiance; oxygen; nutrient concentrations; vegetation characteristics). All models used water flows as inputs, essentially using the hydrological regime as a forcing function. Sometimes the models were coupled to a hydrological model to derive these water flows. Roughly half of the models aimed at modelling greenhouse gas processes in wetlands, while the other half focused more on nutrient (N, P) and organic matter flows and retention. Several of the greenhouse gas models used the model Denitrification-Decomposition (DNDC), which is a process-based biochemistry model of C and N, originally developed for agro-ecosystems (EOS 2017). One paper (Melton et al. 2013) compared 10 models for wetland methane emission.

Table 3.2 lists the vegetation models identified in the survey (Category c, 28 publications). Two main groups of models emerged from this list. One group of models predicted the species composition of the vegetation based on influencing factors such as wetland type, hydrology (e.g. flooding extent) and soil processes. These models predicted the absence or presence of certain species or of vegetation types (species assemblages), or predicted succession patterns. Some of these models were process-based, while others were more descriptive, e.g. by using logistic regression to predict the occurrence of certain vegetation types. One model used satellite imagery to predict the structure of the vegetation (Poulin et al., 2010). The other group of models simulated vegetation biomass, often assuming a mono-specific stand, e.g. of *Phragmites australis* (e.g. Asaeda and Karunaratne, 2000; Soetaert et al., 2004) or *Typha* sp. (Asaeda et al., 2005). These models predicted the growth of the vegetation in terms of belowground and aboveground biomass in relation to irradiance, available nutrients and hydrological conditions.

From the models listed in Tables 3.1 and 3.2, all models that described N and or P processes in detail were listed and numbered 1 to 10 (Table 3.3). In the text below, these numbers were used to refer to the respective models. The types of wetlands covered include lacustrine wetlands (2,5,8), two riverine systems (1,9), one coastal wetland (4) and two wetlands for wastewater treatment (3,7). The Wetlands-DNDC model (6) was more generic, initially developed for forested wetlands (Li et al., 2000), but also applied to other types of wetland vegetation (Zhang et al., 2002). The 'Asaeda models' (10) described processes in common wetland plant species such as *Typha latifolia*, *Typha angustifolia*, *Phragmites australis* and *Phragmites japonica* and can be used for all freshwater systems where these species occur, including constructed wetlands.

All selected models were dynamic simulation models, and 8 out of the 10 models included the growth of biomass. Most models used equations describing photosynthesis and C assimilation to calculate growth (2,5,6,9,10). Three models calculated the growth of biomass in a more simplified way based on growth rates from the literature, carrying capacity, or nutrient limitation (3,4,8). Most models made a distinction between aboveground and belowground biomass and three models did this with species-specific characteristics (2,8,10), while other models had a more general approach (3,5,9) or distinguished trees from undergrowth (6). The WETSAND model (1) did not include biomass at all, while WWQM (7) assumed standard rates for uptake and release of N and P by vegetation, without including biomass as a state variable.

Table 3.1 List of biogeochemical (category B) wetland models identified from literature, including: model type; wetland type and main input and output variables

#	Model type	Wetland type	Input variables	Output variables
B1	Process-based model of P retention in gradient from watershed to lake, with sub-models for hydrology, productivity and P processes	Coastal freshwater	Water flow and water level, irradiance, rainfall and evaporation, plant and plankton productivity, P processes.	P in water and sediment, P retention
B2	Process-based, spatially explicit model of P cycling	Treatment wetland	Water flows, rain, ET, surface water, soil and pore water P concentrations, biomass compartments	Biomass, P concentrations in water, areal extent, P storage
B3	Process-based model of C, N and P	Mangrove	Decay constants of leaf litter, wood turnover and root decomposition constants. Biomass and nutrient concentrations.	Organic matter content (AFDW) and N concentrations (% DW) of mangrove soil profiles (up to 60 cm)
B4	One-dimensional process-based model for diffusion, plant-mediated transport and ebullition of methane	Five wetland sites in North and Central America and Europe	Water table in 170 cm soil profile, soil temperature, NPP, process descriptions	CH4 emissions through diffusion, ebullition, plant-meditated transport
B5	Process-based model for P retention, including hydrology and wetland distribution	Coastal freshwater	Water balance, solar energy, water temperature, plankton /periphyton /macrophyte processes and detrital and sediment system	P retention by coastal wetlands
B6	Process-based model with sub-models for hydrology, primary productivity, sediment, and P.	Constructed wetlands	Water balance, solar energy, water temperature, plankton /periphyton /macrophyte processes and detrital and sediment system	P retention by river riparian wetlands
B7	Wetland-DNDC model (process-based), modified to include management processes like harvesting, burning, water management, fertilization.	Forest wetlands	Hydrology. soil biogeochemistry, vegetation processes and management practices (harvesting, burning, water management, wetland restoration, fertilization, reforestation)	Forest biomass, CO2, CH4 and N2O emissions
B8	Wetland-DNDC model applied to C, water and energy flows in wetlands at two sites	Forest wetlands	Hydrology. soil biogeochemistry, vegetation processes and management practices (harvesting, burning, water management, wetland restoration, fertilization, reforestation)	Forest biomass, CO2, CH4 and N2O emissions
B9	Inversion model of atmospheric transport and chemistry applied to methane emissions 1984-2003	Assessing role of wetlands in global atmospheric CH4 concentrations	Regional atmospheric CH4 fluxes	Contribution of wetlands, forest burning and fossil fuels/other sources to methane flux anomalies
B10	Wetland-DNDC process-based model applied at landscape level by linking to groundwater table dynamics predicted by MIKE-SHE hydrological model	Forest wetlands and landscape	See B7, B9. In this case coupled to MIKE-SHE predictions of water levels as input	See B7, B9. Emissions now estimated regionally for pine flatwood landscape

#	Model type	Wetland type	Input variables	Output variables
B11	Validation of N2O emissions predicted by Wetland-DNDC model using 4-year data from spruce forest system	Forest wetlands	See B7, B9	See B7, B9
B12	McGill Wetland Model (MWM, a peatland C model) based on the Peatland Carbon Simulator (PCARS) and the Canadian Terrestrial Ecosystem Model. Process-based.	Raised ombrotrophic peat bog	Weather, incl. CO2 concentration; soil and vegetation (mosses) characteristics	Net ecosystem production, gross primary production, ecosystem respiration
B13	Process-based simulation model with 4 sub-models: hydrological, N, P, and sediment	Constructed wetlands	Hydrology, TSS, N and P processes. No soil	Effluent TSS, TN, TP concentrations and removal efficiency
B14	Description of DNDC (process-based) model for predicting soil fluxes of N2O, CO2 and CH4 from cropping systems, rice paddies, grazed pastures, forests, and wetlands	Wetlands and agricultural soils	See B7, B9, B10, B11	See B7, B9, B10, B11
B15	Synthesis of global data	Global peat wetlands	N/A	N/A
B16	Peatland C model (McGill Wetland Model MWM), coupled to land surface climate model (wetland version of Canadian Land Surface Scheme, CLASS3W-MWM) applied to a bog and a fen.	Northern peat wetlands	See B12. Coupled to a land surface climate model	Net ecosystem production, gross primary production, ecosystem respiration, C-fluxes
B17	Synthesis of global data	Coastal wetlands	N/A	N/A
B18	10 models for wetland methane emissions were compared in terms of ability to calculate wetland size and CH4 emissions using uniform protocol (WETCHIMP project)	Generic/global	Models used: CLM4Me, DLEM, IAP-RAS, LPJ-Bern, LPJ-WHyMe, LPJ-WSL, ORCHIDEE, SDGVM, UVic-ESCM, UW-VIC	CH4 emission
B19	Simple dynamic simulation model of carbon fluxes between wetlands and atmosphere	Generic/global	Wetland CH4 emission rate, CO2 exchange from atmosphere, 1st-order decay of CH4 in atmosphere	Atmospheric CO2 and CH4 concentrations
B20	Critique of B19	Generic/global	N/A	N/A

B1 (Mitsch and Reeder 1991); B2 (Kadlec 1997); B3 (Chen and Twilley 1999); B4 (Walter and Heimann 2000); B5 (Mitsch and Wang); B6 (Wang and Mitsch 2000); B7 (Li et al. 2004); B8 (Cui et al. 2005); B9 (Bousquet et al. 2006); B10 (Sun et al. 2006); B11 (Lamers et al. 2007); B12 (St Hilaire et al. 2008); B13 (Chavan and Dennett 2008); B14 (Giltrap et al. 2010); B15 (Yu et al. 2010); B16 (Wu et al. 2012); B17 (Pendleton et al. 2012); B18 (Melton et al. 2013); B19 (Mitsch et al. 2013); B20 (Bridgham et al. 2014)

Table 3.2 List of wetland vegetation (category C) models identified from literature, including: model type; wetland type and main input and output variables

#	Model type	Wetland type	Input variables	Output variables
C1	Habitat changes, dynamic. and spatially interacting cells	Coastal	Hydrology, sediment fluxes	Habitats succession
C2	Species succession and abiotic conditions	Created wetland	Soil properties	Plant species associations
C3	Regression model	Riparian	Hydrological Changes	2 functional veg. Types (wooded/herb.)
C4	Structural equation model, abiotic + density effects	Coastal	Biomass, abiotic factors	Plan sp.richn. and biomass
C5	GIS-based (kriging) model of environmental Factors and species patterns	NL wetland area	Ecological factors	Plant species associations
C6	Aeration model of plant roots of different species	General	Water depth, sediment type	Emergent species
C7	Dynamic model of reed growth	Phragmites	Light, temperature	Shoot and root biomass
C8	Architectural, spatially explicit model	Phragmites at high altitudes		Biomass
C9	Vertical growth (sedimentation), based on mass balance	Tidal marshes	Tidal regime, susp. sediment	Tidal flat elevation
C10	Growth and C allocation in Phragmites	Phragmites	Meteo, salinity	Biomass
C11	Dynamic growth model of Typha	Typha	Latitude (light, temp.)	Shoot and root biomass
C12	IBMs, review of approaches and elements	Ecosystems in general		
C13	Empirical models	Lacustrine/riverine wetlands	Hydrology	Wetland vegetation types
C14	Evaluation of dynamic floodplain vegetation model	Floodplains		Vegetation classes
C15	Several regression models	Upland to wetland gradients	Groundwater, hydrology, soil	Vegetation classes a.o.
C16	Rule-based dynamic floodplain vegetation model (CASiMiR)	Riverine wetlands	Hydrology, hydrodynamics	Vegetation classes
C17	Growth model of Phragmites	Phragmites	Sediment conditions, flooding	Shoot and root biomass
C18	Process-based model of life-cycles of 10 abundant plant species	Riparian wetlands	Seasonal flood pattern	10 vegetation types (species)
C19	Structurally dynamic model of two vegetation species	Estuarine wetlands	Sediment deposition, elevation	Extent of Spartina vs Phragmites
C20	Process-based model of C dynamics (Wetland-DNDC)	Wetlands	Soil, hydrology, temp.	Biomass, C fluxes
C21	Conceptual model, 17 components:	Freshwater lacustrine	Nutrient load, sediment load	water level, turb, algae, zoo, fish
C22	A Model Study on the Role of Wetland Zones in Lake Eutrophication and Restoration (PC-Lake)	Lacustrine marshes	P, N, water level	Biomass
C23	PCLake-Marsh module: dynamic model based on Phragmites; aim = model effect on lake water quality	Lacustrine marshes	P, N, water level	Biomass
C24	Floating papyrus model	Papyrus	N, water level	Biomass, N-retention
C25	PEATLAND-VU	Peatlands	Soil, groundwater level, temp.	Carbon, biomass, GHG
C26	Global peat C model (NICE-BGC)	Peatlands	Soil, groundwater level, temp.	Carbon, biomass, GHG
C27	Application and improvement NICE-BGC	Peatlands	Soil, groundwater level, temp.	Carbon, biomass, GHG
C28	CNP in marginal wetlands	Riparian wetlands	CNP, temp, light, oxygen	Biomass, CNP

C1 (Sklar et al. 1985); C2 (Noon 1996); C3 (Toner and Keddy 1997); C4 (Grace and Pugesek 1997); C5 (van Horssen et al. 1999); C6 (Sorrell et al. 2000); C7 (Asaeda and Karunaratne 2000); C8 (Klimeš 2000); C9 (Temmerman et al. 2003); C10 (Soetaert et al. 2004); C11 (Asaeda et al. 2005); C12 (DeAngelis and Mooij 2005); C13 (Hudon et al. 2006); C14 (Benjankar et al. 2010); C15 (Laidig et al. 2010); C16 Benjankar et al. 2011); C17 (Asaeda et al. 2011); C18 (Ye et al. 2013); C19 (Wang et al. 2013); C20 (Zhang et al. 2002); C21 (Chow-Fraser 1998); C22 (Janse et al. 2001); C23 (Sollie et al. 2008); C24 (van Dam et al. 2007); C25 (van Huissteden et al. 2006); C26 (Nakayama 2016); C27 (Nakayama 2017); C28 (van der Peijl and Verhoeven 1999)

Table 3.3 List of selected biogeochemical and vegetation wetland models including nitrogen and or phosphorus processes. Climate classified according to Köppen-Geiger (Peel et al. 2007); + process is included in the model; +/- processes is included, but simplified or lumped with other processes; - process is not included

Selected models, type of wetland	Vegetation/ biomass	Climate (Köppen)*	Nitrogen		Phosphorus		Carbon		Hydrology		Management
			surface water	sediment/soil	surface water	sediment/soil	surface water	sediment/soil	surface	ground	
1 WETSAND, restored permanent freshwater marsh	-	Cfa	+/-	+/-	+/-	+/-	-	-	+	+	+
2 Laurentian Great Lakes, created lacustrine wetlands	*Nelumbo lutea* and various	Dfa/b	-	-	+	+	-	-	+	-	+
3 Biomachine autobiotic model, natural wetland wwt	various	Dfb	-	-	+	+	-	-	+/-	-	-
4 NUMAN, intertidal forested, mangrove	mangroves	Aw	-	+/-	-	+/-	-	+/-	-	-	-
5 PCLake-Marsh, lacustrine marshes	various	Cfb	+	+	+	+	+	+	+	+	+
6 Wetland-DNDC, various	various	various	+	+	-	-	+	+	+	+	+
7 WWQM, constructed wetland for wwt	various	Csb	+	+	+	+	-	-	+	-	+/-
8 Papyrus, permanent freshwater lake wetland	*Cyperus papyrus*	Af	+	+	-	-	-	-	+	-	+
9 Kismeldon, river floodplains	various	Cfb	+	+	+	+	+	+	+	-	+
10 Asaeda macrophytes, emergent macrophytes	*Phragmites* and *Typha*	various	+	+	+	+	+	+	-	-	+/-

1) Kazezyilmaz-Alhan et al. 2007; 2) Mitsch and Reeder 1991; Mitsch and Wang 2000; Wang and Mitsch 2000; 3) Kadlec 1997; 4) Chen and Twilley 1999; 5) Janse et al. 2001; Sollie et al. 2008; 6) Zhang et al, 2002; Sun et al. 2006; Lamers et al. 2007; 7) Chavan and Dennett, 2008; 8) van Dam et al. 2007; 9) van der Peijl and Verhoeven 1999; van der Peijl and Verhoeven 2000; 10) Asaeda and Karunaratne 2000; Asaeda et al. 2002; Asaeda et al. 2005; Asaeda et al. 2008; Asaeda et al. 2011.

* Af = equatorial; Aw = tropical wet and dry; Cfa = humid subtropical; Cfb = marine; Csb = dry summer; Dfa/b humid continental

Eight models were based on wetland data from temperate and subtropical climates with distinct seasons, while only two models (4,8) were based on tropical climates and one model (10) was partly based on a semi-arid subtropical climate. Seven models (1,2,3,4,6,7,10) were completely or partly developed with data from North America, four models (5,6,9,10) simulated European systems, one included Australian data (10) and one was based on data from Africa (8). Five models described seasonal effects as a result of annual climate variability through hydrological and biological, chemical or physical processes (2,5,6,9,10), by including drivers like rainfall, evapotranspiration, temperature, and solar radiation. Three models (1,7,8) achieved seasonal variation through hydrological processes only, and the remaining two models did not include seasonality (3,4).

Eight of the selected models included processes related to N and all of these included organic N and uptake and release of N by vegetation. The models can be roughly divided in two groups. One group consisted of explanatory models that described processes such as N uptake, release, nitrification, denitrification, volatilization and hydrological transport in detail (5,6,8,9,10). The other group used a more empirical approach by lumping processes into simple equations, calibrated and validated with large datasets (1,4,7). All eight models that included N had the objective to better understand or predict the impact of wetlands on water quality.

Equations describing P processes were included in eight of the selected models and six (1,4,5,7,9,10) of these included both N and P. Besides uptake and release by biomass that was included in all models, adsorption and storage of P in sediment were the most common processes. The P models could also be divided in models with a more theoretical (2,3,5,9,10) and models with a more empirical approach (1,4,7). All models were designed to better understand or predict the impact of wetlands on water quality.

Five models (4,5,6,9,10) included C processes. Assimilation and respiration were included in four models and all of these were explanatory (5,6,9,10). Only the DNDC model focussed on C processes specifically and was able to simulate net C emissions. The other models included C processes to simulate vegetative growth and the interactions with N and P decomposition processes.

Eight of the selected models (1,2,4,5,6,7,8,9) included a hydrology component, all included surface water, five pore water, and only three (1,5,6) groundwater. None of the models used a spatial hydrological model, however most models had the potential to be linked to a spatially explicit hydrological (grid) model to generate spatial water quality outputs. All eight models included precipitation, evapo(transpi)ration and surface flow. Other processes were seepage/drainage and sub surface flow. In the WETSAND model, hydrology was the main driver and governed the nutrient related processes through surface and groundwater flows. The other models combined transport with biological and or physical and chemical processes, as described above.

Explicit options to model implications of management regimes on N and P retention were included in eight of the selected models. With most models it was possible to compare different inflows and loading rates or seasonality (e.g. hydrology and temperature regimes). Some models could vary the amount of vegetation (2) or area (5), or introduce harvesting, mowing or grazing regimes (6,8,9). The management options were introduced to show explicitly the impact on water quality (N, P, C), retention or to investigate how the different processes in the nutrient cycles were impacted.

3.4 Conclusion

Development of a model to simulate nutrient retention starts with framing of the complexity of the system and its processes in a proposed conceptual model (Figure 3.1). To develop a generally applicable papyrus model, a simple hydrological model as a forcing factor, for example to introduce seasonality and to transport nutrients, as described in models 5, 6, and

9, is sufficient. For general application a 'square meter model' (a model representing an average square meter of the surface area of the wetland) is suitable and similar to most of the selected models. A square meter model has the potential to be linked to a spatially explicit hydrological model, in which different areas will have different N, P and biomass outputs, due to variation in the hydrology.

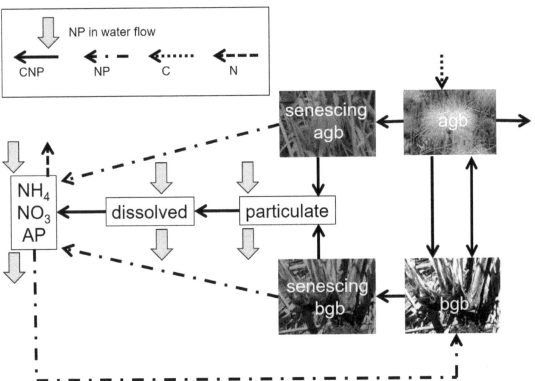

Figure 3.1 Proposed papyrus wetland model set-up with key components and processes with CNP, NP, C and N all mass flows of different forms of C,N and P; NP in water flow=different forms of N and P transported by water flow; NH4=ammonium; NO3=nitrate; AP=available phosphorus; agb=aboveground biomass; bgb=belowground biomass

Table 3.4 Main input variables of the proposed papyrus wetland model

INPUT from other models and the literature			
component	*parameter/process*	*unit*	*data source*
water	water inflow	$m^3\ day^{-1}$	flow data local✦ rivers
climate	rainfall	$m^3\ day^{-1}$	from local✦ meteorological
	evapotranspiration	$m^3\ day^{-1}$	data
	solar radiation	$MJ\ m^{-2}\ day^{-1}$	
N and P	all inputs	$g\ m^{-2}\ day^{-1}$	local✦ data and existing explanatory models
biomass	growth characteristics	$g\ DW\ day^{-1}$	local✦ data and existing
	harvesting	$g\ DW\ day^{-1}$	explanatory models
soil	porosity	%	local✦ data

✦ local means a specific *Cyperus papyrus* dominated wetland with data to parameterize and calibrate the model

An overview of the required inputs and anticipated outputs of the model is presented in Tables 3.4 and 3.5, respectively. The role of biomass is important in the uptake and storage of N and P as well as accumulation of N and P in peat and, therefore, growth (driven by photosynthesis). Consequently mortality, uptake, decay and translocation are also important processes, and these were well described for N in models 5, 6, 8, 9, and 10 and for P in models 2, 3, 5, 9, and 10. Due to the characteristics of *Cyperus papyrus* (Chapter 1) and the role of rhizomes in storage of N and P and reproduction, the distinction between above-and belowground biomass is important. Also important is the interaction between N and P as both nutrients can limit growth and therefore also reduce the uptake of the other. Both the distinction between above- and belowground biomass and combining N and P was best described in models 5, 9 and 10. The Asaeda models (10) also model individual shoots, rhizome and roots separately, while PCLake-Marsh (5) and Kismeldon river floodplains (9) model total biomass. Total biomass seems more suitable for the papyrus model as there may not be sufficient data to model individual shoot (culm) interaction (e.g. shading), and to separate the belowground components roots and rhizomes (Chapter 1).

Table 3.5 Main output variables of the proposed papyrus wetland model

OUTPUT variables

component	parameter/process	unit	relevance
water	volume	m^3	determines concentrations
	inundation levels	m	and saturation level in the wetland
biomass	amount agb and bgb	g DW m^{-2}	determines uptake and storage of N and P
N and P	retention	g m^{-2} yr^{-1}	quantify retention function
	concentration in outflow	g m^{-3}	shows influence wetland on water quality

Based on these requirements of the proposed papyrus wetland model, the selected models (Table 3.3, especially models 5, 9, and 10) provide a sufficient basis to develop the papyrus model. The model could well be used to better quantify the important role of papyrus wetlands for nutrient retention and how this is impacted by harvesting and hydrological dynamics as anthropogenic and natural drivers of change, respectively. The outputs could show the impact of these drivers on water quality outputs, quantify N and P retention and show impact of this on the N to P ratio, which is an important predictor of eutrophication. The model would also allow quantified comparison of the impact of different processes (e.g. peat formation, adsorption, uptake, denitrification) on N and P retention.

Appendix 3.1: List of wetland hydrology models (category a) identified from literature, including: model type; wetland type

#	Model type	Wetland type
A1	Wetland surface water dynamics	Peatland
A2	Diffusion based wetland flow model	Pond
A3	Predictive model for ecological assembly	Riparian wetlands
A4	Wetland groundwater flow for MODFLOW	General
A5	Soil accretionary dynamics and sea level rise	Venice lagoon wetlands
A6	Multidimensional wetland hydrology model, WETLANDS	Cypress pond
A7	Semi distributed stream flow model	Prairie pothole wetlands
A8	Everglades Wetland Hydrodynamic Model (EWHM)	Everglades
A9	Wetting and drying simulation of estuarine processes	Estuary
A10	Groundwater flow model	Pond
A11	Water levels in relation to climate change (WETSIM)	Prairie pothole wetlands
A12	Hydrology/water quality with surface water/groundwater interactions, WETSAND	Restored freshwater
A13	Evapotranspiration from reed marsh	Marsh
A14	Land use change and rainfall impact on sediment export	European loess belt
A15	Hydrologic framework for wetland simulation in climate and earth system models	General
A16	EROSION 2D/3D soil erosion model	General
A17	Influence of wetlands on stream flow hydrology	Riperian wetlands

A1 (Hammer and Kadlec 1986); A2 (Feng and Molz 1997); A3 (Toner and Keddy); A4 (Restrepo et al. 1998); A5 (Day et al. 1999); A6 (Mansell et al. 2000); A7 (Su et al. 2000); A8 (Moustafa and Hamrick 2000); A9 (Ji et al. 2001); A10 (Bravo et al. 2002); A11 (Johnson et al. 2005); A12 (Kazezyilmaz-Alhan et al. 2007); A13 (Zhou and Zhou 2009); A14 (Evrard et al 2010); A15 (Fan and Miquez-Macho 2011); A16 (Schindewolf and Schmidt 2012); A17 (Hughes et al. 2014)

Appendix 3.2: List of integrated wetland models (category d) identified from literature, including: model type; wetland type and main input and output variables

#	Model type	Wetland type
D1	Landscape model, including natural and anthropogenic drivers	Coastal wetlands
D2	General Ecosystem model (GEM), combined hydrological and vegetation model	General
D3	Integrated wetland model: water, nutrients, soils, plants and animals	General/Everglades
D4	Bayesian network model linking hydrology, ecology and livelihood activities	Papyrus
D5	Best Management Practices selection to reduce nutrient loads	Riverine
D6	Landscape modelling and overview of theory and case studies	Various
D7	Conceptual model of wetland degradation and restoration	General
D8	Ecosystem drought recovery	Riverine
D9	Wetland ecosystem services	General
D10	GIS based landscape model	Forested

D1 (Costanza et al. 1990); D2 (Fitz et al. 1996); D3 (Fitz and Hughes 2008); D4 (Van Dam et al 2013); D5 (Yang and Best 2015); D6 (Costanza and Voinov 2003); D7 (Brooks et al. 2005); D8 (Driver et al. 2011); D9 (Feng et al. 2011); D10 (Zhang et al. 2011)

4

A SIMULATION MODEL FOR NITROGEN CYCLING IN NATURAL ROOTED PAPYRUS WETLANDS IN EAST AFRICA[3]

ABSTRACT

Papyrus (*Cyperus papyrus*) wetlands around East African Lakes provide important ecosystem services, including retention of nutrients, to millions of people. To understand the processes contributing to nitrogen retention in the wetland and to evaluate the effects of papyrus harvesting, a dynamic model for carbon and nitrogen cycling in rooted papyrus wetlands was constructed. The model consisted of sub-models for the permanently (P) and seasonally (S) flooded zones and was based on data from a papyrus wetland in Naivasha, Kenya. In each zone, water, nitrogen and carbon flows were calculated based on descriptions of hydrological (river flow, lake level, precipitation, evaporation) and ecological (e.g. photosynthesis, nitrogen uptake, mineralisation, nitrification) processes. Literature data were used for parameterization and calibration. The model simulated realistic concentrations of dissolved nitrogen and papyrus biomass density of papyrus. Daily harvesting up to about 84 (S-zone) and 60 (P-zone) g/m^2*d dry weight reduced the aboveground biomass and increased nitrogen retention (expressed as $(N_{inflow}-N_{outflow})/N_{inflow} * 100\%$) to 38% (S-zone) and 50% (P-zone). A further increase in daily harvesting resulted in collapse of the aboveground biomass. Papyrus biomass, however, recovered fully from annual harvesting of up to 100% of the biomass. The model showed that papyrus re-growth after harvesting is nitrogen-limited in the P-zone.

Key words: Nitrogen, Lake Naivasha, Wetlands, Modelling, Nitrogen retention, Regulating ecosystem services

---

[3] Published as:

Hes EMA, Niu R, van Dam AA (2014) A simulation model for nitrogen cycling in natural rooted papyrus wetlands in East Africa. Wetl Ecol Manag 22(2):157-176

4.1 INTRODUCTION

Wetlands dominated by *Cyperus papyrus* are important ecosystems in East and Central Africa because they provide important ecosystem functions and services for millions of people, including provisioning services for building, crafts and fuel, water, food and medicinal herbs (Geheb & Binns 1997; Kipkemboi *et al.* 2007b; Mwakubo & Obare 2009; Kabumbuli & Kiwazi 2009), and regulating ecosystem services such as sediment and nutrient retention or flood regulation (Mwanuzi *et al.* 2003; Loiselle *et al.* 2008). They provide a habitat for mammals, birds and fish (Gichuki & Gichuki 1992; Maclean *et al.* 2006; van Dam *et al.* 2011). Papyrus vegetation can be rooted in the sediment, or through the action of wind and waves, become detached to form floating mats (Azza *et al.* 2006). Papyrus wetlands thus often show zones with distinct hydrological character: a floating outer fringe; a permanently flooded zone, in which the soil is saturated year-round and *C. papyrus* is the dominant species; and a seasonally flooded zone, which can be dry during part of the year (Denny 1984).

In decision making about wetlands, provisioning services often get priority over regulating services despite the fact that regulating services generally have the higher monetary value (Stuip *et al.* 2002; Emerton 2005). Regulating services are difficult to value and appreciate because of a lack of knowledge of the underlying processes. Lacking clear commodity prices; their value must be estimated using indirect methods (De Groot *et al.* 2006). Many papyrus wetlands are under pressure, driven by population growth and economic development. Structural changes to the wetland like conversion to cropland, construction of irrigation and drainage canals, fish traps and hippo ditches affect the ecological functioning of the system. Livelihoods activities enhance the provisioning services of the wetlands but reduce the regulating services, thus disturbing the balance among ecosystem services important for sustainable management (TEEB 2010; Maltby and Acreman 2011). Better quantitative understanding of regulating services of wetlands, and of their dynamics under the influence of natural and anthropogenic pressures is required (Carpenter *et al.* 2009).

Nutrient retention in wetlands depends on hydrology, atmospheric flux of nitrogen, phosphorus adsorption capacity of the soil, and export of nutrients through vegetation harvesting. Uptake of nutrients is related directly to the growth rate of the plants, with papyrus capable of absorbing large quantities of nutrients from soil and water during the exponential growth phase. Nutrient uptake is reduced when growth rate decreases because of suboptimal soil wetness or maturity of the vegetation (van Dam *et al.* 2007). Nutrients contained in vegetation are released during senescence. Harvesting of plants affects nutrient balances, in two ways. First, nutrients are simply removed. Second, reduced biomass of the vegetation stimulates re-growth and nutrient uptake. Denitrification and biological nitrogen fixation influence the nitrogen balance of the wetland directly, but little is known about these processes in papyrus wetlands (van Dam *et al.* 2011).

Retaining nutrients in wetlands reduces enrichment of downstream water bodies. Eutrophication in the East African lakes has a direct impact on the local economies as water ways get blocked by excessive growth of water hyacinth and fish catches reduce due to algal blooms. The impact of papyrus on downstream nitrogen concentrations is higher than on phosphorus concentrations as papyrus contains more N (the median N:P ratio of papyrus in ten different sites was 19.85; Gaudet 1975). Therefore a model to simulate nitrogen in a papyrus wetland to compare the effects of wetness and harvesting on outflow concentrations will be useful for determining trade-offs between provisioning services (harvesting) and regulating services (nutrient retention).

The overall aim of this study is to better understand the effects of harvesting and hydrology on nitrogen retention in rooted papyrus wetlands. Specific objectives were: (1) to construct a dynamic simulation model of rooted papyrus growth in both seasonally and permanently flooded zones; (2) to parameterize and calibrate the model with a dataset from Lake Naivasha in Kenya; (3) to assess the effects of hydrological conditions and harvesting on nitrogen retention.

4.2 METHODS

4.2.1 System description and model development

Lake Naivasha papyrus wetland is one of the best studied papyrus systems (e.g. Gaudet 1979; Jones and Muthuri 1997; Becht *et al.* 2006; Boar 2006). Owing to water level fluctuations and cultivation, the area covered by *C. papyrus* declined from around 50 to 5-10 km^2 between 1960 and 2000 (Hickley *et al.* 2004). Studies between 1993 and 2001 showed that mean dry weight (DW) biomass was 6.9 g/m^2, with approximate culm densities of 30 culms/m^2 and plant height of up to 4 m (Boar 2006). The north swamp in Naivasha is located at the mouth of the Malewa River which floods the wetland for short periods during the rainy season (Boar and Harper 2002). At the lakeward side, the wetland is flooded by lake water.

Based on this situation, a conceptual model for a rooted papyrus wetland was constructed (Figure 4.1). River water floods the seasonal (S) zone and then the permanent (P) zone. Depending on the water level of the lake, the P-zone discharges into or receives backflow from the lake. Because of the slope of the S-zone, the surface water flows towards the P-zone and there is no standing water in the S-zone. Precipitation, evapotranspiration and groundwater flows affect the water depth in the wetland. The model assumes a monospecific stand of *C. papyrus*.

Several models of the role of vegetation in the nutrient cycling of wetlands have been developed. Van der Peijl and Verhoeven (1999) modelled a sloping wetland at Kismeldon Meadows, Devon, southwestern England. The model describes the carbon, nitrogen and phosphorus dynamics, their interactions in a riverine wetland with a vegetation dominated by *Molinia caerulea* and how these are affected by soil wetness and temperature. Based on descriptions of basic processes such as photosynthesis, nutrient uptake and decomposition the model describes the nutrient cycles and plant growth reasonably well. Van Dam *et al.* (2007) constructed a model of a floating papyrus wetland in Jinja, Uganda. This model described the flows of nitrogen through the papyrus vegetation and various compartments of detritus, and predicted the effects of vegetation harvesting on the water quality under the floating mat. These two models were used as the basis for the new papyrus model. A new model was needed as the existing models focus on different plant species and climate (Kismeldon Meadows) or on floating papyrus (Jinja) and therefore do not sufficiently cover the processes and components of a rooted papyrus wetland.

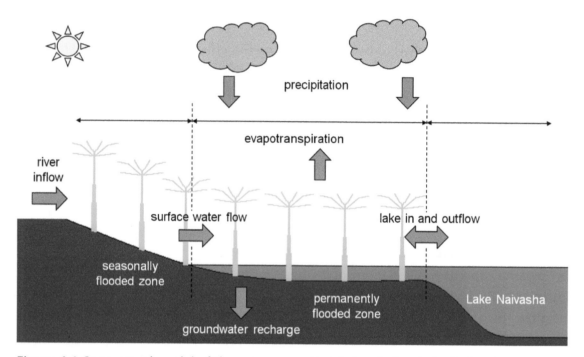

Figure 4.1 Conceptual model of the papyrus wetlands bordering Lake Naivasha

4.2.2 EQUATIONS, VARIABLES AND CONSTANTS

Table 4.1 lists the state variables used and Table 4.2 presents all processes with equations. Key variables and processes are presented below in the model description, in which numbers behind variables or processes refer to the state variables in Table 4.1 or to equations in Table 4.2. A complete list of variables and constants is provided in Appendix 4.1. The model was built using STELLA 9.1.4 (High Performance Systems, Hanover, NH) and run for a period of 5 years with rectangular (Euler) integration and a time step of 0.0625 days (1.5 hours) for one square meter of wetland. A complete listing of the equation layer of the Stella model can be found in Appendix 4.2.

4.2.3 MODEL ASSUMPTIONS AND IMPLEMENTATION

The model assumes an even distribution of water in the wetland without channels or preferential flow. Because of the slope in the S-zone, surface water flows towards the P-zone and there are no pools in the S-zone. All papyrus plants are rooted, and floating papyrus mats at the edge of the lake are not considered. Phosphorus, sulphur and other elements needed for growth are assumed not to be limiting and not included in the model. Growth is limited when there is no water to provide ammonium or nitrate. The model does not include nitrogen fixation, nitrogen deposition and ammonia volatilization. Nitrogen fixation in papyrus wetlands is poorly quantified (Mwaura and Widdowson 1992; Gichuki *et al.* 2005). Ammonia volatilization is not expected to play a big role as the pH in papyrus wetlands is low (Azza *et al.* 2000). Nitrogen (wet and dry) deposition in East Africa is estimated at 0.5 g N /m^2*yr (Dentener *et al.* 2006), which is less than 0.5% of the total N inflow used in the model simulations.

Table 4.1 State variables

variable	description	#	initial value	unit	source
WaterS	water level seasonally flooded wetland	1	0.2	m^3/m^2	a
WaterP	water level permanently flooded wetland	2	0.5	m^3/m^2	a
CAGBS	C aboveground biomass seasonally flooded	3	1853	g C/m^2	b
CBGBS	C belowground biomass seasonally flooded	4	1570	g C/m^2	b
CDAGBS	C dead AGB seasonally flooded	5	335	g C/m^2	c
CDBGBS	C dead BGB seasonally flooded	6	284	g C/m^2	c
POCS	particulate organic carbon seasonally flooded	7	20	g C/m^2	a
CAGBP	C aboveground biomass permanently flooded	8	1853	g C/m^2	b
CBGBP	C belowground biomass permanently flooded	9	1570	g C/m^2	b
CDAGBP	C dead AGB permanently flooded	10	335	g C/m^2	c
CDBGBP	C dead BGB permanently flooded	11	284	g C/m^2	c
POCSP	POC surface water permanently flooded	12	20	g C/m^2	a
POCPP	POC pore water permanently flooded	13	20	g C/m^2	a
NAGBS	N aboveground biomass seasonally flooded	14	44	g N/m^2	d
NBGBS	N belowground biomass seasonally flooded	15	31	g N/m^2	d
NDAGBS	N dead AGB seasonally flooded	16	7.9	g N/m^2	c
NDBGBS	N dead BGB seasonally flooded	17	5.6	g N/m^2	c
PONS	particulate organic nitrogen seasonally flooded	18	0.9	g N/m^2	e
DONS	dissolved organic nitrogen seasonally flooded	19	1.1	g N/m^2	e
NO3S	nitrate seasonally flooded	20	0.05	g N/m^2	e
NH4S	ammonium seasonally flooded	21	0.5	g N/m^2	e
NH4AS	ammonium adsorbed seasonally flooded	22	5	g N/m^2	a
NAGBP	N aboveground biomass permanently flooded	23	44	g N/m^2	d
NBGBP	N belowground biomass permanently flooded	24	31	g N/m^2	d
NDAGBP	N dead AGB permanently flooded	25	7.9	g N/m^2	c
NDBGBP	N dead BGB permanently flooded	26	5.6	g N/m^2	c
PONSP	PON surface water permanently flooded	27	0.9	g N/m^2	e
PONPP	PON pore water permanently flooded	28	0.9	g N/m^2	e
DONSP	DON surface water permanently flooded	29	1.1	g N/m^2	e
DONPP	DON pore water permanently flooded	30	1.1	g N/m^2	e
NH4SP	ammonium surface water permanently flooded	31	0.5	g N/m^2	e
NH4PP	ammonium pore water permanently flooded	32	0.5	g N/m^2	e
NO3SP	nitrate surface water permanently flooded	33	0.05	g N/m^2	e
NO3PP	nitrate pore water permanently flooded	34	0.05	g N/m^2	e
NH4AP	ammonium adsorbed permanently flooded	35	0.5	g N/m^2	a

a = estimate; b = average of Boar et al. (1999) Jones and Humphries (2002) Boar (2006); c = calculated, see Appendix 4.1; d = average of Boar *et al.* (1999) Boar (2006); e = Muthuri and Jones (1997)

Table 4.2 Rate variables (processes) in the model

process	description	equation	#
lake_Inflow	lake inflow	IF(surfw_P<threshold__swd_P) THEN(lake_inflow_rate) ELSE(0)	36
outflow	outflow of P wetland	IF(surfw_P>threshold__swd_P) THEN(surface_water_flow+prec_P-evap_P-recharge_P) ELSE(0)	37
prec_P	precipitation	rainfall_rate*0.001	38
evap_P	evaporation	evaporation_rate*0.001	39
recharge_P	groundwater recharge	frout_P*porew_free_P	40
surface_water_flow	surface water flow from S to P wetland	froff_S*surfw_S	41
evap_S	evaporation	evaporation_rate*0.001	42
prec_S	precipitation	rainfall_rate*0.001	43
inflow_S	inflow S wetland	river_inflow_rate/area_S	44
recharge_S	groundwater recharge	frout_S*porew_free_S	45
CAGBS_harvest_D	daily harvesting AGB	(PULSE(harvest_in_g_C, harvest_day-1,harvest_interval))*harvest_in_S_yes_or_no	46
CAGBS_harvest_A	annual harvesting AGB	(PULSE(CAGBS*harvest_%, harvest_day-1,harvest_interval))*harvest_in_S_yes_or_no	47
CAGBS_respiration	respiration AGB	CAGBS*maintenance_coefficient+CAGBS_assimilation*growth_coefficient	48
CAGBS_assimilation	CO_2 uptake AGB	max_assimilation_constant*CAGBS*limit_radiance*(limit_N_S/0.9)*((max_AGB_biomass*perc_C_in_AGB-CAGBS)/max_AGB_biomass*perc_C_in_AGB)	49
CAGBS_death	death of AGB	CAGBS*CAGB_death_constant	50
CDAGBS_leach	leaching from DAGB	CDAGB_leach_constant*CAGBS_death	51
CDAGBS_frag	fragmentation of DAGB	CDAGBS*CDAGB_frag_constant	52
POCS_hydrolysis	hydrolysis of POC	POCS*POC_hydrolysis_constant	53
CDBGBS_frag	fragmentation of DBGB	CDBGBS*CDBGB_frag_constant	54
CDBGBS_leach	leaching from DBGB	CDBGB_leach_constant*CBGBS_death	55
CBGBS_death	death of BGB	CBGBS*CBGB_death_constant	56
CBGBS_respiration	respiration BGB	CBGBS*maintenance_coefficient+C_trans_S*growth__coefficient	57
C_trans_S	translocation	(CAGBS-CBGBS*C_AGB_to_BGB_optimal_ratio)/(1+C_AGB_to_BGB_optimal_ratio)	58 59
CAGBP_harvest_D	daily harvesting AGB	PULSE(harvest_in_g_C, harvest_day-1,harvest_interval)	60
CAGBP_harvest_A	batch harvesting AGB	PULSE(CAGBP*harvest_%, harvest_day-1,harvest_interval)	61
CAGBP_respiration	respiration AGB	CAGBP*maintenance_coefficient+CAGBP_assimilation*growth_coefficient	62
CAGBP_assimilation	CO_2 uptake AGB	max_assimilation_constant*CAGBP*limit_radiance*limit_NP_P*((max_AGB_biomass*perc_C_in_AGB-CAGBP)/(max_AGB_biomass*perc_C_in_AGB))	63
CAGBP_death	death of AGB	CAGBP*CAGB_death_constant	64
CDAGBP_leach	leaching from DAGB	CDAGB_leach_constant*CAGBP_death	65
CDAGBP_frag	fragmentation of DAGB	CDAGBP*CDAGB_frag_constant	66
POCSP_hydrolysis	hydrolysis of POC in surface water	POCSP*POC_hydrolysis_constant	67
POCPP_hydrolysis	hydrolysis of POC in pore water	POCPP*POC_hydrolysis_constant	68
POCSP_settling	settling of POC	POCSP*POC_settling_rate	69
CDBGBP_frag	fragmentation of DBGB	CDBGBP*CDBGB__frag_constant	70

CDBGBP_leach	leaching from DBGB	CDBGB_leach_constant*CBGBP_death	71
CBGBP_death	death of BGB	CBGBP*CBGB_death_constant	72
CBGBP_respiration	respiration BGB	CBGBP*maintenance_coefficient+C_trans_P*growth__coefficient	73
C_trans_P	translocation	(CAGBP-CBGBP*C_AGB_to_BGB_optimal_ratio)/ (1+C_AGB_to_BGB_optimal_ratio)	74
NAGBS_harvest_D	daily harvesting AGB	CAGBS_harvest_D/CN_AGBS_ratio	75
NAGBS_harvest_B	batch harvesting AGB	IF(CN_AGBS_ratio>0)THEN(CAGBS_harvest_B/CN_AGBS_ratio) ELSE(0)	76
NAGBS_death	death of AGB	CAGBS_death*(1-N_retrans_constant)/CN_AGBS_ratio	77
NDAGBS_frag	fragmentation of DAGB	CDAGBS_frag/CN_DAGBS_ratio	78
PONS_inflow	inflow of particulate organic nitrogen	PONS_load	79
PONS_hydrolysis	hydrolysis of PON	PONS*PON_hydrolysis_constant	80
DONS_inflow	inflow of dissolved organic nitrogen	DONS_load	81
DONS_outflow	outflow of DON	surface_water_flow*conc_DONS	82
DONS_mineral	mineralisation	DONS*K_mineral	83
DONS_recharge	groundwater recharge	conc_DONS*recharge_S	84
PONS_outflow	PON outflow	surface_water_flow*conc_PONS	85
NDBGBS_frag	fragmentation	CDBGBS_frag/CN_DBGBS_ratio	86
PONS_recharge	groundwater recharge	conc_PONS*recharge_S	87
NH4S_outflow	NH4 outflow	surface_water_flow*conc_NH4S	88
NH4S_adsorption	adsorption	IF(conc_NH4S>5)THEN(K_NH4_adsorption*NH4S)ELSE (-K_NH4_adsorption*NH4AS))	89
NH4S_recharge	groundwater recharge	conc_NH4S*recharge_S	90
NDBGBS_leach	leaching from DBGB	CDBGBS_leach/CN_DBGBS_ratio	91
NH4S_inflow	NH4 inflow	NH4S_load	92
NDAGBS_leach	leaching from DAGB	CDAGBS_leach/CN_DAGBS_ratio	93
NH4S_uptake	uptake by papyrus	max_NH4_uptake*N_papyrus_S* (1-N_papyrus_S/N_max_papyrus)*limit_NH4S	94
nitri_S	nitrification	K_nitri*mode_S*NH4S	95
NO3S_outflow	NO3 outflow	surface_water_flow*conc_NO3S	96
NO3S_recharge	groundwater recharge	conc_NO3S*recharge_S	97
denitri_S	denitrification	K_denitri*NO3S*(1-mode_S)	98
NO3S_inflow	NO3 inflow	NO3S_load	99
NO3S_uptake	uptake by papyrus	max_NO3_uptake*N_papyrus_S*(1-N_papyrus_S/N_max__papyrus)* limit_NO3S	100
NBGBS_death	death of BGB	CBGBS_death/CN_BGBS_ratio	101
N_trans_S	translocation	(NAGBS-NBGBS*C_AGB_to__BGB_ratio_S)/(C_AGB_to_BGB_ratio_S+1)	102
N_retrans_S	retranslocation	CAGBS_death*N_retrans_constant/CN_AGBS_ratio	103
NAGBP_harvest_D	daily harvesting AGB	CAGBP_harvest_D/CN_AGBP_ratio	104
NAGBP_harvest_B	batch harvesting AGB	IF(CN_AGBP_ratio>0)THEN(CAGBP_harvest_B/CN_AGBP_ratio) ELSE(0)	105
NAGBP_death	death of AGB	CAGBP_death*(1-N_retrans_constant)/CN_AGBP_ratio	106
NDAGBP_leach	leaching from DABG	CDAGBP_leach/CN_DAGBP_ratio	107
NDAGBP_frag	fragmentation	CDAGBP_frag/CN_DAGBP_ratio	108
PONSP_lake_in	PON inflow lake	lake_Inflow*conc_PON__lake_inflow	109
PONSP_settling	settling of PON	PONSP*PON_settling_constant	110
PONSP_surf_in	PON inflow surface water	surface_water_flow*conc_PONS	111

PONSP_outflow	PON outflow	outflow*conc_PONSP	112
PONSP_hydrolysis	hydrolysis of PON in surface water	PONSP*PON_hydrolysis_constant	113
DON_w_lake_in	DON inflow lake	lake_Inflow*conc_DON__lake_inflow	114
DON_w_surf_in	DON inflow surface water	surface_water_flow*conc_DONS	115
DONP_diffusion	DON diffusion between surface and pore water	IF(surfw_P>0) THEN(K_DON_diffusion*((DONSP/surfw_P)-(DONPP/soil_depth_P))/((surfw_P+soil_depth_P)/2)) ELSE(0)	116
DONSP_outflow	DON outflow	outflow*conc_DONSP	117
DONSP_mineral	mineralisation in surface water	K_mineral*DONSP	118
NH4_lake_in	NH$_4$ inflow lake	lake_Inflow*conc_NH4_lake_inflow	119
NH4S_in	NH$_4$ inflow surface water	surface_water_flow*conc_NH4S	120
NH4P_diffusion	NH$_4$ diffusion between surface and pore water	IF(surfw_P>0) THEN(K_NH4_diffusion*((NH4SP/surfw_P)-(NH4PP/soil_depth_P))/((surfw_P+soil_depth_P)/2)) ELSE(0)	121
NH4SP_outflow	NH$_4$ outflow	outflow*conc_NH4SP	122
nitri_SP	nitrification surface water	K_nitri*NH4SP	123
NH4P_adsorption	adsorption	(IF(conc_NH4PP>5)THEN(NH4PP*K_NH4_adsorption)ELSE(-2*K_NH4_adsorption*NH4AP))	124
NO3S_in	NO$_3$ inflow surface water	surface_water_flow*conc_NO3S	125
NO3_lake_in	NO$_3$ inflow lake	lake_Inflow*conc_NO3__lake_inflow	126
NO3SP_outflow	NO$_3$ outflow	outflow*conc_NO3SP	127
NO3P_diffusion	NO$_3$ diffusion between surface and pore water	IF(surfw_P>0) THEN(K_NO3_diffusion*((NO3SP/surfw_P)-(NO3PP/soil_depth_P))/((surfw_P+soil_depth_P)/2)) ELSE(0)	128
denitri_PP	denitrification	K_denitri*NO3PP*(1-mode_P)	129
NO3P_recharge	groundwater recharge	conc_NO3PP*recharge_P	130
NO3_uptake_P	NO$_3$ uptake by papyrus	max_NO3_uptake*N_papyrus_P*(1-N_papyrus_P/N_max_papyrus)*limit_NO3PP	131
nitrifi_PP	nitrification pore water	K_nitri*mode_P*NH4PP	132
NH4P_recharge	groundwater recharge	conc_NH4PP*recharge_P	133
NDBGBP_leach	leaching from DBGB	CDBGBP_leach/CN_DBGBP_ratio	134
NH4_uptake_P	NH$_4$ uptake by papyrus	max_NH4_uptake*N_papyrus_P*(1-N_papyrus_P/N_max_papyrus)*limit_NH4PP	135
DONPP_mineral	mineralisation	K_mineral*DONPP	136
PONPP_hydrolysis	hydrolysis of PON in pore water	PONPP*PON_hydrolysis__constant	137
DONP_recharge	groundwater recharge	conc_DONPP*recharge_P	138
PONP_recharge	groundwater recharge	conc_PONPP*recharge_P	139
NDBGBP_frag	fragmentation	CDBGBP_frag/CN_DBGBP_ratio	140
NBGBP_death	death of BGB	CBGBP_death/CN_BGBP_ratio	141
N_trans_P	translocation	(NAGBP-NBGBP*C_AGB_to_BGB_ratio_P)/(C_AGB_to_BGB_ratio_P+1)	142
N_retrans_P	retranslocation	CAGBP_death*N_retrans_constant/CN_AGBP_ratio	143

4.2.4 Parameterization and calibration

The model was parameterized and calibrated with literature data available for Lake Naivasha (Appendix 4.1). When values from Lake Naivasha were not available, data from other East African papyrus wetlands were used. For parameters that were never studied or measured in papyrus wetlands, literature values from other wetland types were used or values estimated

(Table 4.1 and Appendix 4.1). Seasonal variability in the Lake Naivasha wetland was described using monthly averages from the period 1970-1982 for irradiance (Muthuri *et al.* 1989), evaporation and precipitation from the period 1974-1976 (Gaudet 1979) and river inflow. River inflow had different values for each month, based on the flow regime of the Malewa river (Gaudet 1979) and was calibrated to achieve realistic flow rates and nitrogen concentrations. River flow followed a similar seasonal pattern as rainfall (Figure 4.5), with a main rainy season in the months March, April and May and a short rainy season in December. Inflow in the dry season (0.09 m/day), was about half of the inflow at the peak of the rainy season (0.19 m/day). The peak rainfall (0.19 m/day), was equal to the maximum inflow rate. During most of the year rainfall was lower than inflow. Evaporation varied throughout the year between 0.10 and 0.15 m/day. There was an outflow of the P-zone during about 130 days per year with a peak of 0.33 m/day. The rest of the year there was a backflow from the lake of between 0 and 0.12 m/day. This hydrological regime was repeated annually.

4.2.5 HYDROLOGY SUB-MODEL

The hydrology sub-model (Figure 4.2) calculates the water level in S and P-zones (as the volume of water per m² of surface area) based on river discharge into the S-zone, surface water flow from the S-zone into the P-zone, outflow and backflow between the P-zone and the lake, and precipitation, evaporation and groundwater recharge in both S-zone and P-zone. River discharge, precipitation and evaporation were based on observed data (Gaudet 1979). Groundwater recharge was modelled with first order equations (equations 40 and 45). If the soil is saturated, 1% of the pore space above the water filled porosity is recharged, resulting in a maximum recharge of 0.6 mm per day.

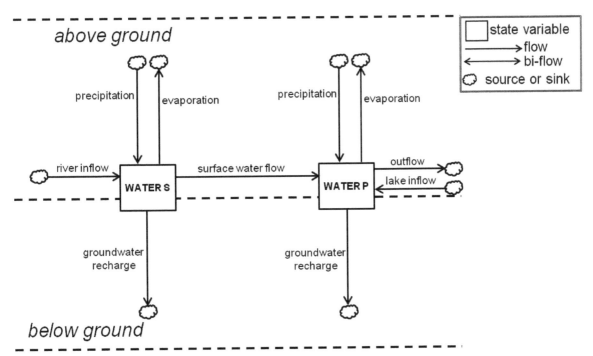

Figure 4.2 Conceptual diagram of the hydrology sub-model (WATER S = water level in the seasonal zone; WATER P = water level in permanent zone)

The term MODE (Appendix 4.1) controls the availability of oxygen in the wetland. This is important as oxygen availability controls nitrification and denitrification rates. Based on the amount of water and on soil volume and porosity, the water filled porosity is calculated and compared with the water filled porosity at field capacity. Below field capacity, conditions are aerobic (MODE = 1). When the soil is fully saturated or flooded, conditions are anaerobic (MODE = 0). Between field capacity and saturation, the value of MODE is related linearly to the proportion of pore space filled (MODE between 0 and 1).

4.2.6 CARBON SUB-MODELS

Papyrus biomass was modelled as carbon in aboveground (culms and umbels) and belowground (rhizome and roots) biomass (Figure 4.3). Carbon in the aboveground biomass (CAGB) results from assimilation (photosynthesis) and translocation of carbon from the rhizome. Assimilation (equations 49, 63) was modelled as a logistic model depending on aboveground biomass, and limited by irradiance and the availability of nitrate and ammonium (both Monod-type equations):

$$\text{Assimilation}=\text{max\_assimilation\_constant}\cdot\text{CAGB}\cdot\frac{\text{radiance}}{(\text{radiance} + \text{K\_radiance})}\cdot\frac{\text{limit\_N}}{0.9}\cdot\frac{(\text{max\_AGB\_biomass}\cdot\text{perc\_C\_in\_AGB}-\text{CAGB})}{\text{max\_AGB\_biomass}\cdot\text{perc\_C\_in\_AGB}}$$

in which max_assimilation_constant is the maximum relative assimilation rate (day^{-1}), CAGB is carbon in the aboveground biomass (g C/m^2), radiance is irradiance (MJ/m^2 day), K_radiance is the half saturation constant of irradiance for assimilation (MJ/m^2 day), limit_N is limiting factor of N for carbon assimilation (dimensionless; see Appendix 4.1), the factor 0.9 ensures that maximum growth limitation can be reached (van der Peijl and Verhoeven 1999), max_AGB_biomass is the maximum AGB biomass (g/m^2), and perc_C_in AGB is the carbon content (dry weight) of aboveground biomass (%).

It was assumed that whenever the nitrogen concentration in papyrus was below the minimum needed for assimilation (0.0016 g N/ g DW; van der Peijl and Verhoeven 1999), there was no growth (limit_N_P and limit N_S, Appendix 4.1) so a Monod-type function with a cut-off was used.

Translocation of carbon between aboveground and belowground biomass (equations 59, 74) was based on an assumed optimal ratio between aboveground and belowground carbon (C_AGB_to_BGB_optimal_ratio, Appendix 4.1), of 1.2 (Boar *et al.* 1999; Jones and Humphries 2002). Whenever carbon ratio differed from this optimal ratio, carbon was translocated to restore the optimum:

$$\text{Translocation}=\frac{(\text{CAGB}-\text{CBGB}\cdot\text{optimal\_C\_AGB}/\text{C\_BGB})}{(1+\text{optimal\_C\_AGB}/\text{C\_BGB})}$$

in which CAGB and CBGB are carbon in aboveground and belowground biomass, respectively (g C/m^2), and optimal_C_AGB/C_BGB is the optimal ratio between carbon in aboveground and belowground biomass (g C/m^2).

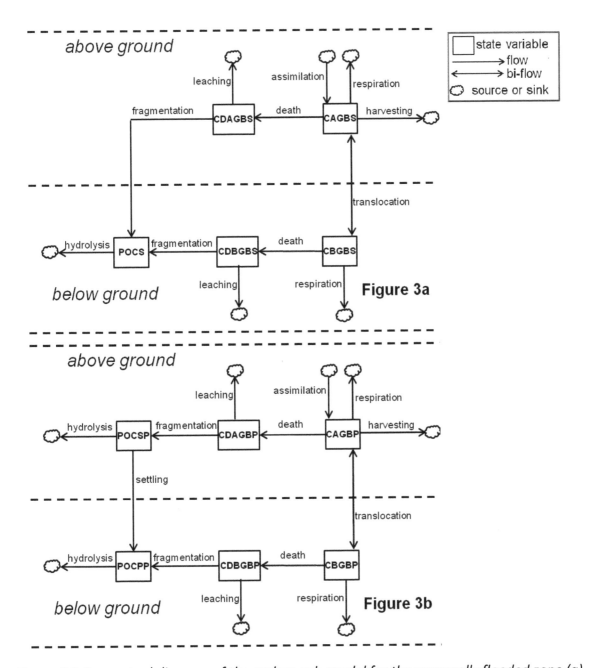

Figure 4.3 Conceptual diagram of the carbon sub-model for the seasonally flooded zone (a) and permanently flooded zone (b) (names of state variables are explained in Table 4.1)

Other processes leading to a reduction in CAGB were respiration (the sum of maintenance and growth respiration; equations 48 and 57) and harvesting (equations 46,47,60 and 61). Respiration was calculated assuming that maintenance respiration is proportional to biomass and that growth respiration was proportional to assimilation and translocation, as:

$$Respiration = maint\_coeff\_AGB \cdot CAGB + maint\_coeff\_BGB \cdot CBGB + growth\_coeff\_AGB \cdot assimilation + growth\_coeff\_BGB \cdot translocation$$

in which maint_coeff_AGB and maint_coeff_BGB are the maintenance respiration coefficients for aboveground and belowground biomass, respectively (day^{-1}), and growth_coeff_AGB and growth_coeff_BGB are the respiration coefficients for assimilation and translocation, respectively (-).

Both CAGB and CBGB were subject to death (flow to carbon in dead biomass), fragmentation (aboveground and belowground dead biomass being converted to particulate organic carbon in the sediment) and hydrolysis. All carbon processes were defined separately for the S and P-zones (Figures 4.3A and 4.3B). In the S-zone, because of the short residence time (about 2 days) of the surface water, hydrolysis of particulate organic carbon was assumed to take place in the pore water only.

4.2.7 NITROGEN SUB-MODELS

The nitrogen sub-models (Figure 4.4) express the same aboveground and belowground live and dead biomass compartments as the carbon model in terms of nitrogen. Added to these are the main components of the nitrogen cycle in the wetland. Nitrate and ammonium, originating from river and lake inflow as well as from the microbial breakdown of dead papyrus, are taken up by the belowground biomass, and then passed on to the aboveground biomass by translocation. Uptake rates (equations 94, 100, 131 and 135) were related directly to CO_2-assimilation assuming a fixed C/N ratio in the papyrus biomass (Table 4.2). Nitrogen in dead biomass passes through fragmentation, hydrolysis, mineralisation and nitrification. Nitrification (equations 95, 123 and 132) and denitrification (equations 98 and 129) were moderated by oxygen availability through the factor MODE. Ammonium adsorption to the soil (equations 89 and 124) takes place when the NH_4 concentration is above 5 g N/m^2.

All nitrogen processes were defined separately for the S and P-zones. In the S-zone model, only belowground processes taking place in the pore water were modelled because surface water in this zone has a short residence time. Exchange of N between the aboveground and belowground layers in the P-zone takes place through diffusion, driven by concentration differences of soluble compounds (nitrate, ammonium and dissolved organic nitrogen) and settling of particles (particulate organic nitrogen).

The uptake of ammonium and nitrate by papyrus depends on the carrying capacity for papyrus (max_AGB_biomass) and is limited by the concentration of ammonium and nitrate, respectively. This limitation was modelled with a Monod-type equation:

$$\text{NH4S\_uptake}=\text{max\_NH4\_uptake}\cdot\text{N\_papyrus\_S}\cdot\left(1-\frac{\text{N\_papyrus\_S}}{\text{N\_max\_papyrus}}\right)\cdot\frac{\text{conc\_NH4S}}{(\text{conc\_NH4S}+\text{K\_NH4})}$$

in which max_NH4_uptake is the maximum uptake rate of ammonium by papyrus (day^{-1}), N_papyrus_S is the amount of nitrogen in above- and belowground biomass in the S zone (g/m^2), N_max_papyrus is the maximum amount of nitrogen stored in papyrus biomass (g/m^2), conc_NH4S is the ammonium concentration in the S zone (g N/m^3), and K_NH4 is a half saturation constant (g N/m^3).

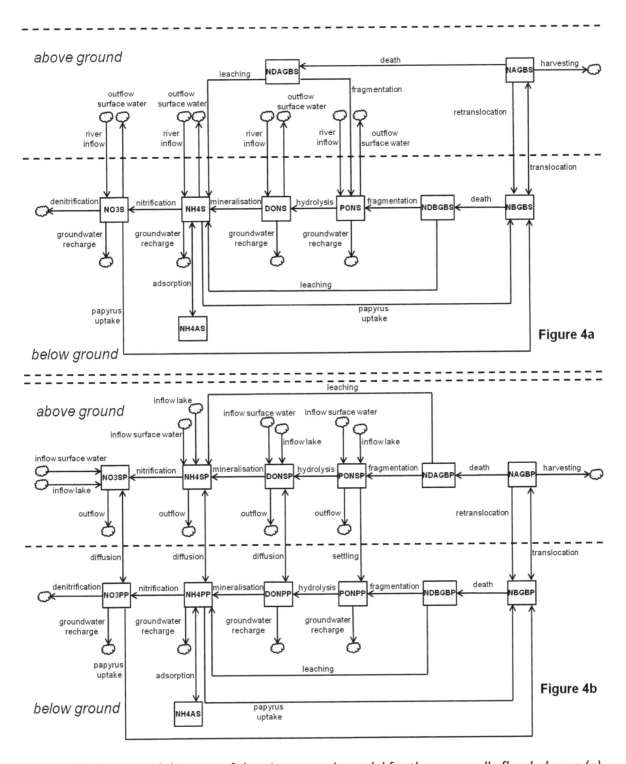

Figure 4.4 Conceptual diagram of the nitrogen sub-model for the seasonally flooded zone (a) and permanently flooded zone (b) (names of state variables are explained in Table 4.1)

N_max_papyrus was calculated based on literature values for nitrogen content in both above and belowground biomass, 0.013 g N/g DW and 0.008 g N/g DW respectively (Appendix 4.1) and the maximum papyrus density in literature, 8118 g DW/m^2 (Muthuri *et al.* 1989 and Jones and Muthuri 1997). Equations for nitrate uptake in the S zone (100) and ammonium and nitrate uptake in the P zone (131 and 135) are listed in Table 4.2.

4.2.8 NITROGEN RETENTION AND HARVESTING SCENARIOS

Nitrogen retention (g N m^{-2} y^{-1}) was calculated as $(N_{inflow}-N_{outflow})/N_{inflow}$ * 100% for the S and P-zones separately, in which N_{inflow} for the S-zone was the amount of nitrogen carried with inflow_S (44) in a year, and N_{inflow} for the P wetland was the sum of the nitrogen in lake_inflow (36) and surface_water_flow (41). The $N_{outflow}$ of the S wetland was the sum of surface_water_flow (41) and recharge_S (45), and the $N_{outflow}$ of the P wetland was outflow (37) plus recharge_P (40). Retention was calculated over the 5th year when the system was observed to be stable.

Seven harvesting scenarios were defined: no harvest, daily 40 (daily harvest of 40 grams dry weight of aboveground papyrus biomass per square meter), daily 95 (95 grams of aboveground papyrus biomass), daily 120 (120 grams of aboveground papyrus biomass), annual 33 (33% of the aboveground dry weight biomass, harvested annually at the 90th day of the year), annual 67 (67% of the aboveground biomass) and annual 100 (100% of the aboveground biomass).

4.3 RESULTS

4.3.1 WATER DEPTH, WATER QUALITY AND PAPYRUS GROWTH

The water level in the S-zone (Figure 4.6) showed two peaks coinciding with the two rainy seasons. The water level in the P-zone was always above 0.3 meter, with the lowest level just before the start of the main rainy season. In the S-zone the wetland fell completely dry at the beginning and the end of the year. Without harvesting, aboveground papyrus (Figure 4.6) fluctuated between 3480 and 3485 g DW/m^2 within a year, with an initial value of 3489 g DW/m^2 (Appendix 4.1). Belowground biomass fluctuated between 3890 and 3893 g DW/m^2, after starting at a value of 3928 g DW/m^2. Total biomass was around 7375 g DW/m^2 (corresponding to 80 g N/m^2) for both S-zone (Figure 4.6) and P-zone. Nitrogen in aboveground biomass (Figure 4.6) was higher (41-43 g N/m^2) than in belowground biomass (35-37 g N/m^2). At the beginning of the year, before the start of the rainy season, the nitrogen in both aboveground and belowground biomass in the S-zone decreased and then increased again after the onset of the rains. Most of the year there was more nitrogen stored in biomass in the S-zone than in the P-zone. Nitrate and ammonium levels in the surface water of the P-zone (Figure 4.6) increased during the period of the year when there was no outflow (Figure 4.5) to 1.7 and 25 g N/m^3, respectively. During the period with outflow from the P-zone to the lake, the concentrations dropped to 0.5 g N/m^3 for nitrate and 3 g N/m^3 for ammonium.

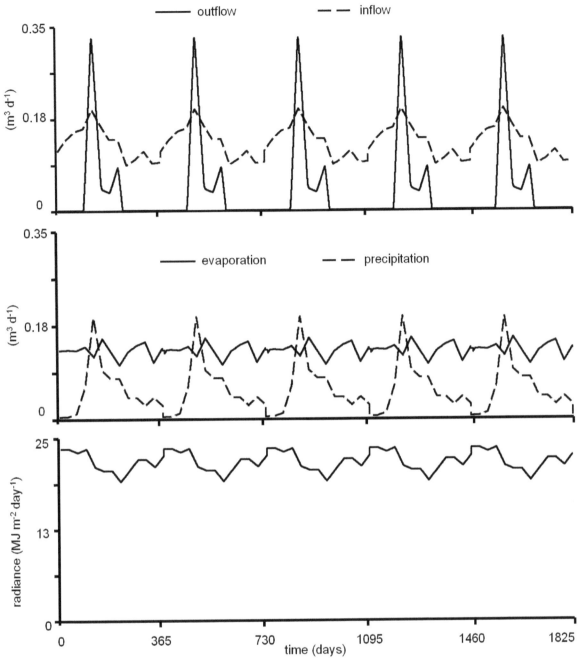

Figure 4.5 River inflow in the S zone, outflow to the lake from the P zone, evaporation, precipitation and radiance

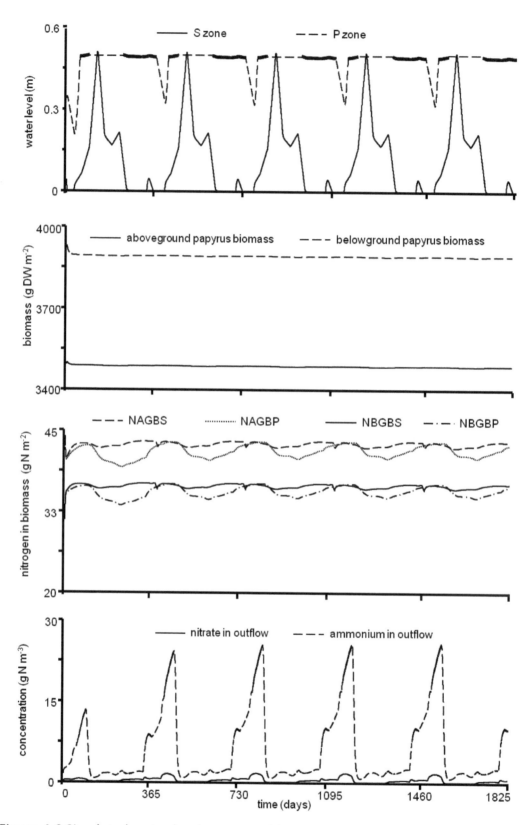

Figure 4.6 Simulated water levels, papyrus biomass, nitrogen in papyrus biomass (NAGBS = Nitrogen in aboveground biomass in S zone; NAGBP = nitrogen in aboveground biomass in P zone; NBGBS = nitrogen in belowground biomass in S zone; NBGBP = nitrogen in belowground biomass in P zone) and nitrate and ammonium concentrations in the outflow of the P zone

4.3.2 EFFECT OF HARVESTING ON ABOVEGROUND BIOMASS

Figure 4.7 shows the nitrogen (in g N/m^2) in the aboveground parts of the papyrus in the S and P-zones with the seven harvesting scenarios. The amount of nitrogen stored in living aboveground papyrus decreased with increasing daily harvesting rates and stabilized over time. This implies that papyrus remained present in the wetland, but with a lower density than without harvesting. At a harvesting rate of 95 g/m^2*d, the system collapsed and the papyrus disappeared. According to the simulations this happened after two and a half year in the S zone and within a year in the P zone. The papyrus in the P zone collapsed earlier due to low nitrate and ammonium concentrations in the root zone (max 0.1 an 0.2 g N/m^3 respectively). With annual harvesting in the S-zone, the papyrus recovered to its original density even if all aboveground papyrus was harvested. The recovery after harvesting of more than 67% took longer than half a year. For the P-zone, recovery took longer than in the S-zone again due to low nitrate and ammonium concentrations. The papyrus was not able recover to its original density if more than 67% was harvested each year.

4.3.3 EFFECTS OF HARVESTING ON OUTFLOW CONCENTRATIONS OF NITRATE AND AMMONIUM

The nitrate and ammonium concentrations in the surface water of the P-zone (Figure 4.8), which is also the concentration of the outflow, increased during the period without outflow due to accumulation. During the period with outflow the concentrations dropped. Daily harvesting (daily 40) reduced both nitrate and ammonium concentrations, however when the harvesting rate increased (daily 95) the concentrations were even higher than without harvesting, with no dead biomass accumulation (Table 4.3). With annual harvesting, both nitrate and ammonium concentrations were lower as the amount harvested increased. Ammonium concentrations were higher than nitrate concentrations throughout.

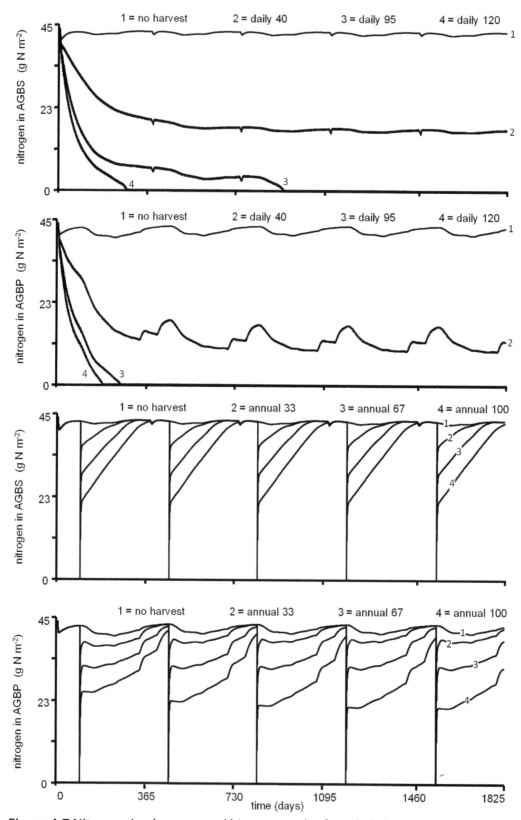

Figure 4.7 Nitrogen in aboveground biomass under four daily harvesting scenarios: no harvest, daily 40 (40 g DW of papyrus), daily 95 (95 g DW of papyrus) and daily 120 (120 g DW of papyrus) and four annual harvesting scenarios: no harvest, annual 33 (33 % of aboveground biomass), annual 67 (67 % of aboveground biomass) and annual 100 (100 % of aboveground biomass) in the S zone and P zone respectively

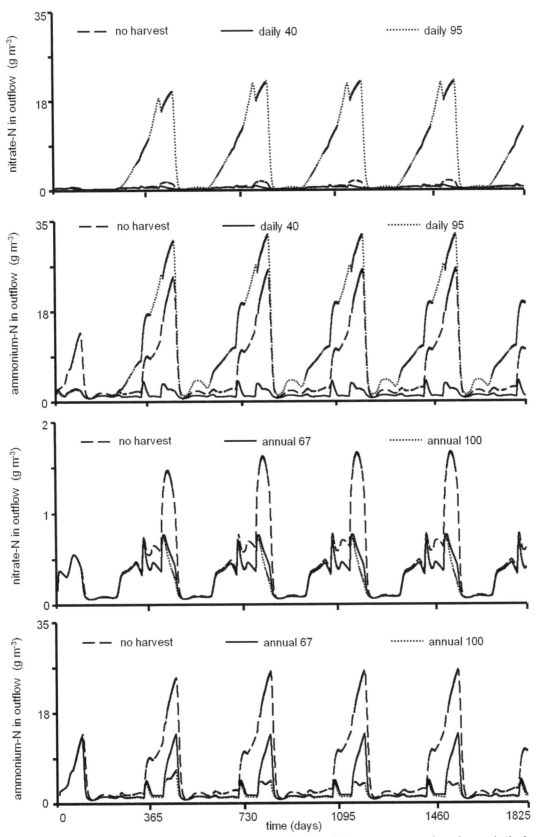

Figure 4.8 Nitrate and ammonium in the outflow of the P zone under three daily harvesting scenarios: no harvest, daily 40 (40 g DW of papyrus) and daily 95 (95 g DW of papyrus) and three annual harvesting scenarios: no harvest, annual 67 (67 % of aboveground biomass) and annual 100 (100 % of aboveground biomass)

4.3.4 EFFECTS OF HARVESTING ON NITROGEN RETENTION

Figure 4.9 compares nitrogen retention in the S-zone and the P-zone. Under the daily harvesting scenario (daily harvesting of aboveground biomass), harvesting increased N-retention from 13 to 50% for the S-zone and from 14 to 38% for the P-zone. This is due to re-growth of the papyrus after harvesting. The difference between the S and P-zones is caused by nitrogen limitation in the P-zone. When daily harvesting is increased further, there is a sudden decrease in retention. This tipping point occurs between papyrus harvesting rates of 84 and 96 g /m²*d in the S-zone and between 60 and 72 g/m²*d in the P-zone. Beyond this point the amount harvested was bigger than the re-growth and therefore the papyrus is not able to recover and there was no aboveground biomass.

*Table 4.3 Nitrogen budget for S zone and P zone over the fifth year without harvesting and with daily harvesting of 40, 95 or 120 g/m2*d (DW). Numbers are flows in g N/m2*y (without harvesting) and in % of total N input into the wetland (all scenarios), for both S-zone and P-zone. N$_{accum}$ is the net difference in a compartment between start and end of the year. N$_{inflow}$ = N$_{in}$ (PON+DON+DIN) and N$_{outflow}$ = N$_{out}$ (PON+DON+DIN)*

		N-flow									
	Compartment	no harvest		no harvest		daily 40		daily 95		daily 120	
		S	P	S	P	S	P	S	P	S	P
		(g N m^{-2} y^{-1})		(% of N$_{in}$)		(% of N$_{in}$)		(% of N$_{in}$)		(% of N$_{in}$)	
N$_{in}$	PON	36.5	33.0	22	20	22	20	22	20	22	20
	DON	36.5	57.7	22	35	22	35	22	35	22	35
	DIN	91.3	75.5	56	45	56	45	56	45	56	45
N$_{accum}$	AGB	0	0	0	0	0	0	0	0	0	0
	BGB	0	0	0	0	0	0	0	0	0	0
	DAGB	19.0	18.7	12	11	4	3	-1	0	0	0
	DBGB	2.0	2.0	1	1	0	0	0	0	0	0
	PON	0.1	0.2	0	0	0	0	0	0	0	0
	DON	0.1	1.5	0	1	0	1	0	0	0	0
	DIN	0	0	0	0	0	0	0	0	0	0
	adsorbed	0	0	0	0	0	0	0	0	0	0
N$_{out}$	PON	32.6	16.3	20	10	16	8	14	8	13	8
	DON	59.0	90.1	36	54	33	48	32	46	31	46
	DIN	51.0	37.2	31	22	2	7	53	43	53	43
	denitrification	0.4	0.1	0	0	0	0	2	3	2	3
	harvesting	0	0	0	0	45	33	0	0	0	0
N$_{inflow}$		164.3	166.2	100	100	100	100	100	100	100	100
N$_{outflow}$		142.6	143.6	87	86	51	64	98	97	98	97
N-retention	Total	21.7	22.6	13	14	49	36	2	3	2	3

Nutrient retention without harvesting is mainly the accumulation of dead papyrus biomass (Table 4.3). Once the aboveground biomass is zero there was no longer accumulation of dead biomass and retention was only caused by denitrification (Table 4.3). Retention in the P-zone without aboveground biomass is higher than in the S zone (Figure 4.9) because of denitrification (Table 4.3). The anaerobic conditions required for denitrification occur for a longer period of the year in the P-zone.

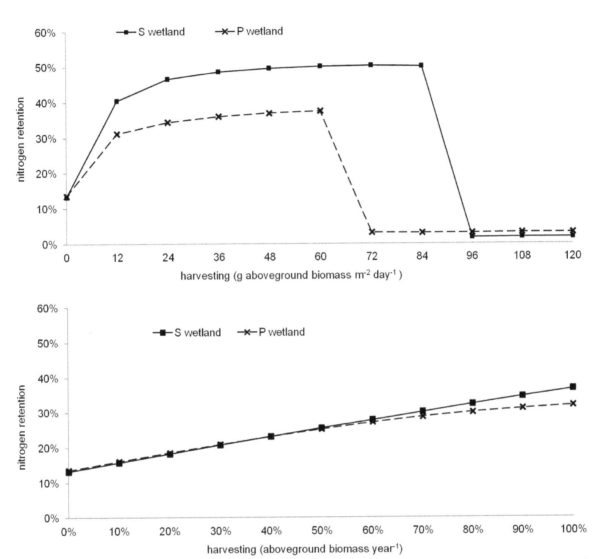

Figure 4.9 Nitrogen retention in S zone and P zone under two harvesting scenarios: daily harvesting (top) and annual harvest at highest biomass density (bottom)

Figure 4.9 also shows the N-retention under annual harvesting, expressed as a percentage of the total *aboveground* biomass. In the S-zone, N-retention increased linearly, indicating that all the harvested papyrus grew back before the next harvest. In the P-zone, there was also a linear increase up to about 40% of harvesting. For these harvesting rates, N-retention was slightly higher (about 1.5 g N/m^2*y) compared with the S-zone because of accumulation of dissolved organic nitrogen (Table 4.3). Above 40% harvesting, nitrogen became limiting in the P-zone, leading to a smaller increase in retention. The papyrus was not able to grow back to the same density within a year.

4.4 DISCUSSION AND CONCLUSION

Simulated papyrus biomass density (7375 g DW/m^2) was comparable with literature values of 6945 g DW/m^2 given in Boar (2006) and of 7775 g DW/m^2 in Jones and Muthuri (1997). Nitrogen in aboveground and belowground biomass was 41-43 and 35-37 g N/m^2, respectively, which compares well with values of 44 for aboveground and 31 g N/m^2 for belowground nitrogen measured in the field (Boar *et al.* 1999; Boar 2006). Nitrate and ammonium concentrations of 0.5 and 3 g N/m^3 in the outflow were also realistic (Gaudet 1979; Cózar *et al.* 2007). Field observations show that it takes 6-12 months for aboveground biomass to grow back from rhizomes and roots. In the model (S-zone), this was about 9 months (Figure 4.7). The dataset used for calibration was compiled from different studies in Naivasha spread over an extended time period (1989 - 2006). While the model simulated the Naivasha papyrus system well, calibration and validation with field data from other papyrus wetlands is needed to confirm that the model predicts papyrus growth correctly within a range of hydrological and biogeochemical conditions.

Because of differences in hydrology, the nitrogen processes were modelled differently in the S and P-zones (Figures 4.2 and 4.4). Due to rapid flow of surface water to the P-zone and its short residence time (2 days) in relation to the rates of the processes in the surface water, in the S-zone only pore water processes were considered. In the P-zone, the residence time of surface water was longer and therefore both surface and pore water processes were included. During parts of the year with backflow from the lake instead of outflow (Figure 4.4), there was an accumulation of dissolved nitrogen and, therefore, elevated concentrations in the P-zone surface water (Figure 4.6). Nitrogen concentrations increased even further when the water level dropped temporarily because the evaporation was higher than the sum of all water inputs (Figure 4.2).

Where possible, parameter values from published studies on papyrus vegetation of Lake Naivasha were used (see also Appendix 4.1). Some constants were taken from the "parent" models (van der Peijl and Verhoeven 1999; van Dam *et al.* 2007; Table 4.2 and Appendix 4.1). Further work on the model will include a sensitivity analysis to identify the parameters that influence the model outputs most and could be the focus of additional field or laboratory research.

Without harvesting, N-retention was between 13 and 14 % of N-input, equivalent to approximately 22 g N/m^2*y, similar to 21.5 g N/m^2*y estimated for a floating papyrus wetland (van Dam *et al.* 2007). The model results suggest that the major part of this nitrogen (19 g/m^2*y) accumulates in dead aboveground biomass (DAGB) and the remainder in dead belowground biomass (DBGB), particulate and dissolved organic nitrogen (Table 4.3). Accumulated detritus may eventually be removed from the wetland by floods (van Dam *et al.* 2007). Nitrogen is removed permanently by denitrification, estimated in the model at 0.1-0.4 g N/m^2*y as a result of low nitrate concentrations. Based on model simulations, Mwanuzi *et* al. (2003) also concluded negligible denitrification in papyrus wetlands. Field measurements on denitrification in papyrus wetlands that can be used to verify model results are scarce.

The effect of harvesting on N-retention is positive. Harvesting reduces papyrus biomass and stimulates re-growth thus increasing nitrogen uptake. As the papyrus in the S-zone can grow back within a year, annual batch harvests, even up to 100% of the *aboveground* papyrus, increase N-retention (Figure 4.9B). According to the model simulations, the effect of daily harvesting on N-retention is positive up to a harvesting rate of about 90 and 70 g DW/m^2 for the S-zone and P-zone, respectively (Figure 4.9A). The faster growth in the S-zone predicted by this mono-specific model does not take into account competition from other plant species that may occur under dry conditions (Rongoei *et al.* 2013). Above harvesting rates of about 90 and 70 g DW/m^2 for S and P-zone respectively, the system collapses (Figure 4.7 and 4.9) and N-retention is reduced to denitrification (Table 4.3).

The effects of hydrology and harvesting on N-retention show a strong interaction. Without harvesting, the difference between N-retention in the S-zone and the P-zone was small. With harvesting rates below 90 g/m^2*d aboveground biomass (Table 4.3 and Figure 4.9), N-retention in the S-zone was higher than in the P-zone because of low nitrogen concentrations in the P-zone which limit re-growth (Figure 4.8). These low concentrations occurred during the wet season, when water flowed from the P-zone to the lake and nitrogen could not accumulate to replenish the nitrogen taken up by the papyrus. Therefore when there was a net outflow to the lake, papyrus densities decreased with harvesting in the P zone (Figure 4.5). With harvesting rates above 90 g/m^2*d, papyrus collapses and N-retention was only due to denitrification. Denitrification rates in the P-zone were higher because of anaerobic conditions throughout the year, whereas the S-zone is anaerobic only part of the year.

While harvesting contributes to N-retention at the wetland scale as it exports nitrogen from the wetland, at the basin scale it only contributes to N-retention if the harvested material is not decomposed within the same basin. As harvested papyrus is often used locally for construction, handicrafts and fish traps, it could be argued that the nitrogen is not exported from the basin.

For the model to contribute to better understanding of nutrient regulation function of papyrus wetlands the model needs further development. Gaudet (1975) showed that the N content of different papyrus organs (root, rhizome, culm and umbel) varies from 0.86 in the culm to 2.67 %DW in the rhizome. Consequently, the N content of a young, growing papyrus stand can be expected to change with the change in development stage of the plants. The current model, with a constant N content, could be improved by incorporating an allometric relationship between N content and biomass. A further improvement is the incorporation of phosphorus in the model. This would allow the estimation of P-retention in the papyrus wetland. P-retention is affected by harvesting (P content: 0.024-0.099 % DW in 10 different sites in Africa; Gaudet 1975), however, adsorption to the soil is more important in P retention (Kelderman *et al.* 2007). Including phosphorus will also allow the evaluation of N:P ratios in the plants and the effects of stoichiometry on nutrient cycling. As plants take up nutrients in proportions that maintain their optimal element ratio (Elser *et al.* 2010), variations in growth stages and vegetation harvesting may affect the N:P ratio of the outflow. This may have implications for the nutrient regulation function of papyrus wetlands and the water quality in the adjoining lake.

Appendix 4.1: Variables and constants of the papyrus model

name	description	equation or value	unit	source
area_S	area of the S wetland	1000	m^2	artificial
CAGB_death_constant	death rate of aboveground biomass	0.0057	day^{-1}	van der Peijl and Verhoeven, 1999
CBGB_death_constant	death rate of belowground biomass	0.0014	day^{-1}	van der Peijl and Verhoeven, 1999
CDAGB_frag_constant	fragmentation rate of aboveground biomass	2.76×10^{-4}	day^{-1}	van der Peijl and Verhoeven, 1999
CDABG_leach_constant	maximum fraction leached of aboveground biomass	0.432	-	van der Peijl and Verhoeven, 1999
CDBGB_frag_constant	fragmentation rate of belowground biomass	8.34×10^{-4}	day^{-1}	van der Peijl and Verhoeven, 1999
CDBGB_leach_constant	maximum fraction leached of belowground biomass	0.486	-	van der Peijl and Verhoeven, 1999
CN_AGBP_ratio	C to N ratio in aboveground biomass in P wetland	CAGBP/NAGBP	g C/g N	
CN_AGBS_ratio	C to N ratio in aboveground biomass in S wetland	CAGBS/NAGBS	g C/g N	
CN_BGBP_ratio	C to N ratio in belowground biomass in P wetland	CBGBP/NBGBP	g C/g N	
CN_BGBS_ratio	C to N ratio in belowground biomass in S wetland	CBGBS/NBGBS	g C/g N	
CN_DAGBP_ratio	C to N ratio in dead aboveground biomass in P wetland	CDAGBP/NDAGBP	g C/g N	

name	description	equation or value	unit	source
CN_DAGBS_ratio	C to N ratio in dead aboveground biomass in S wetland	CDAGBS/NDAGBS	g C/g N	
CN_DBGBP_ratio	C to N ratio in dead belowground biomass in P wetland	CDBGBP/NDBGBP	g C/g N	
CN_DBGBS_ratio	C to N ratio in dead belowground biomass in S wetland	CDBGBS/NDBGBS	g C/g N	
conc_DONPP	concentration of dissolved organic nitrogen in pore water of P wetland	DONPP/pore_water_P	g/m^3	
conc_DONS	concentration of dissolved organic nitrogen in S wetland	IF(WaterS>0)THEN (DONS/WaterS) ELSE(0)	g/m^3	
conc_DONSP	concentration of dissolved organic nitrogen in surface water of P wetland	DONSP/surfw_P	g/m^3	
conc_DONS_inflow	concentration of dissolved organic nitrogen in inflow of S wetland	DONS_load/inflow_S	g/m^3	
conc_DON_lake_inflow	concentration of dissolved organic nitrogen in lake	0.065	g/m^3	estimate
conc_NH4PP	concentration of ammonium in pore water of P wetland	NH4PP/pore_water_P	g/m^3	
conc_NH4S	concentration of ammonium in S wetland	IF(WaterS>0) THEN(NH4S/WaterS) ELSE(0)	g/m^3	
conc_NH4SP	concentration of ammonium in surface water of P wetland	NH4SP/surfw_P	g/m^3	

name	description	equation or value	unit	source
conc_NH4S_inflow	concentration of ammonium in inflow S wetland	NH4S_load/inflow_S	g/m³	
conc_NH4_lake_inflow	concentration of ammonium in lake	0.8	g/m³	estimate
conc_NO3PP	concentration of nitrate in pore water of P wetland	NO3PP/pore_water_P	g/m³	
conc_NO3S	concentration of nitrate in S wetland	IF(WaterS>0) THEN(NO3S/WaterS) ELSE(0)	g/m³	
conc_NO3SP	concentration of nitrate in surface water of P wetland	NO3SP/surfw_P	g/m³	
conc_NO3S_inflow	concentration of nitrate in inflow of S wetland	NO3S_load/inflow_S	g/m³	
conc_NO3_lake_inflow	concentration of nitrate in lake	0.5	g/m³	estimate
conc_PONPP	concentration of particulate organic nitrogen in pore water of P wetland	PONPP/pore_water_P	g/m³	
conc_PONS	concentration of particulate organic nitrogen in S wetland	IF(WaterS>0)THEN (PONS/WaterS)ELSE(0)	g/m³	
conc_PONSP	concentration of particulate organic nitrogen in surface water of P wetland	PONSP/surfw_P	g/m³	
conc_PONS_inflow	concentration of particulate organic nitrogen in inflow of S wetland	PONS_load/inflow_S	g/m³	
conc_PON_lake_inflow	concentration of particulate organic nitrogen in lake	0.065	g/m³	estimate
C_AGB_to_BGB_optimal _ratio	optimal C AGB to C BGB ratio	initial_C_AGB/initial_ C_BGB	-	

name	description	equation or value	unit	source
C_AGB_to_BGB_ratio_P	C AGB to C BGB ratio in P wetland	CAGBP/CBGBP	-	
C_AGB_to_BGB_ratio_S	C AGB to C BGB ratio in S wetland	CAGBS/CBGBS		
C_conc_in_plant	concentration of C in plant	(initial_C_AGB+initial_C_BGB)/ (initial_AGB+initial_BGB)	g C/g DW	
DONS_load	dissolved organic nitrogen entering the S wetland	0.1	g N/day	estimate
evaporation_rate	evaporation rate in Naivasha region	COUNTER(0,365)	mm/day	Gaudet, 1978
froff_S	fraction of water that flows from S wetland to P wetland	0.5	day^{-1}	estimate
frout_P	fraction of water recharged to groundwater in P wetland	0.01	day^{-1}	estimate
frout_S	fraction of water recharged to groundwater in S wetland	0.01	day^{-1}	estimate
growth_coeff_AGB	coefficient of respiration represented by growth component, proportional to photosynthesis for AGB	0.3	day^{-1}	Bachelet et al., 1989
growth_coeff_BGB	coefficient of respiration represented by growth component, proportional to photosynthesis for BGB	0.2	day^{-1}	Bachelet et al., 1989
harvest_%	percentage of AGB harvested	value between 0 and 100	%	
harvest_day	first day of harvest	value between 1-365	-	

name	description	equation or value	unit	source
harvest_interval	number of days between each harvest		-	
harvest_in_g_C	harvest expressed in g of carbon	harvest_in_g_papyrus * perc_C_in_AGB	g C/m²	
harvest_in_g_papyrus	harvest expressed in g of papyrus DW	variable	g DW/m²	
harvest_in_S_yes_or_no	factor determining if harvesting in S wetland takes place or not	0 or 1	-	
initial_AGB	initial weight of aboveground biomass	3489	g DW/m²	average of Muthuri *et al.,* 1989. Jones and Muthuri, 1997 Boar, 2006
initial_BGB	initial weight of belowground biomass	3928	g DW/m²	average of Jones and Muthuri, 1997 Boar, 2006
initial_C_AGB	initial carbon in aboveground biomass	1853	g C/m²	average of Boar *et al.,* 1999 Jones and Humphries, 2002 Boar, 2006
initial_C_BGB	initial carbon in belowground biomass	1570	g C/m²	average of Boar *et al.,* 1999 Jones and Humphries, 2002 Boar, 2006
initial_C_DAGB	initial carbon in death aboveground biomass	initial_dead_papyrus_biomass* ratio_AGB_total_biomass* perc_C_in_AGB	g C/m²	
initial_C_DBGB	initial carbon in death aboveground biomass	initial_dead_papyrus_biomass* ratio_BGB_total_biomass* perc_C_in_BGB	g C/m²	
initial_dead_papyrus_biomass	initial weight of dead papyrus biomass	1340	g DW/m²	estimate based on Boar, 2006

name	description	equation or value	unit	source
initial_N_AGB	initial nitrogen in aboveground biomass	44	g N/m^2	average of Boar *et al.*, 1999 Boar, 2006
initial_N_BGB	initial nitrogen in belowground biomass	31	g N/m^2	Boar *et al.*, 1999 Boar, 2006
initial_N_DAGB	initial nitrogen in dead aboveground biomass	initial_dead_papyrus_biomass* ratio_AGB_total_biomass* perc_N_in_AGB	g N/m^2	
initial_N_DBGB	initial nitrogen in dead belowground biomass	initial_dead_papyrus_biomass* ratio_BGB_total_biomass* perc_N_in_BGB	g N/m^2	
initial_papyrus_biomass	initial weight of papyrus biomass	initial_AGB+initial_BGB	g DW/m^2	
kmN	Concentration of N in the plant at which limiting factor is 0.5	N_conc_minimum+((N_conc_optimal-N_conc_minimum)/9)	g N/g DW	
K_denitri	denitrification rate	0.01	day^{-1}	estimate
K_DON_diffusion	diffusion rate constant for dissolved organic nitrogen	0.05	m^2/day	van Dam *et al.*, 2007
K_mineral	mineralization rate	2*10^{-4}	day^{-1}	van Dam *et al.*, 2007
K_NH4	half saturation constant	0.7	g N/m^3	van Dam *et al.*, 2007
K_NH4_adsorption	adsorption rate	0.01	day^{-1}	estimate
K_NH4_diffusion	diffusion rate constant for ammonium	0.05	m^2/day	van Dam *et al.*, 2007
K_nitri	nitrification rate	0.005	day^{-1}	estimate
K_NO3	half saturation constant	0.1	g N/m^3	van Dam *et al.*, 2007
K_NO3_diffusion	diffusion rate constant for nitrate	0.05	m^2/day	van Dam *et al.*, 2007
K_radiance	half saturation constant	1	MJ/m^2*day	estimate
lake_inflow_rate	lake inflow rate in P wetland	0.12	m^3/m^2*day	estimate

name	description	equation or value	unit	source
limit_NH4PP	limitation factor for uptake of ammonium in P wetland	conc_NH4PP/ (conc_NH4PP+K_NH4)	-	
limit_NH4S	limitation factor for uptake of ammonium in S wetland	conc_NH4S/ (conc_NH4S+K_NH4)	-	
limit_NO3PP	limitation factor for uptake of nitrate in P wetland	conc_NO3PP/ (conc_NO3PP+K_NO3)	-	
limit_NO3S	limitation factor for uptake of nitrate in S wetland	conc_NO3S/ (conc_NO3S+K_NO3)	-	
limit_N_P	N limiting factor for carbon assimilation in P wetland	IF((NP_conc_in_plant-N_conc_minimum)/((kmN-N_conc_minimum)+(NP_conc_in_plant-N_conc_minimum))<(0.9-NP_conc_in_plant))A ND(0.9-NP_conc_in_plant)>N _conc_minimum THEN(NP_conc_in_pla nt-N_conc_minimum)/((kmN-N_conc_minimum)+(NP_conc_in_plant-N_conc_minimum))EL SE IF((NP_conc_in_plant-N_conc_minimum)/((kmN-N_conc_minimum)+(NP_conc_in_plant-N_conc_minimum))>= (0.9-NP_conc_in_plant))A ND((0.9-NP_conc_in_plant)>N _conc_minimum) THEN(0.9)ELSE(0)	-	van der Peijl and Verhoeven, 1999
limit_N_S	N limiting factor for carbon	IF((NS_conc_in_plant-N_conc_minimum)/((	-	van der Peijl and Verhoeven, 1999

name	description	equation or value	unit	source
	assimilation in S wetland	kmN-N_conc_minimum)+(NS_conc_in_plant-N_conc_minimum))<(0.9-NS_conc_in_plant))AND(0.9-NS_conc_in_plant)>N_conc_minimum THEN(NS_conc_in_plant-N_conc_minimum)/((kmN-N_conc_minimum)+(NS_conc_in_plant-N_conc_minimum))ELSE IF((NS_conc_in_plant-N_conc_minimum)/((kmN-N_conc_minimum)+(NS_conc_in_plant-N_conc_minimum))>=(0.9-NS_conc_in_plant))AND((0.9-NS_conc_in_plant)>N_conc_minimum) THEN(0.9)ELSE(0)		
limit_radiance	radiance limitation of carbon assimilation	radiance/(radiance+K_radiance)	-	
maint_coeff_AGB	maintenance coefficient for AGB	0.02	day^{-1}	Bachelet *et al.*, 1989
maint_coeff_BGB	maintenance coefficient for BGB	0.002	day^{-1}	Bachelet *et al.*, 1989
max_AGB_biomass	maximum ABG biomass	max_papyrus_biomass* ratio_AGB_total_biomass	g/m^2	
max_assimilation_constant	maximum assimilation of carbon by papyrus	0.5	day^{-1}	estimate
max_NH4_uptake	maximum uptake of ammonium	0.05	day^{-1}	van Dam *et al.*, 2007

name	description	equation or value	unit	source
max_NO3_uptake	maximum uptake of nitrate	0.05	day^{-1}	van Dam *et al.*, 2007
max_papyrus_biomass	maximum papyrus biomass	8118	g DW/m^2	Muthuri *et al.*, 1989 and Jones and Muthuri, 1997
mode_P	controlling factor for oxygen availability in P wetland	IF(wfp_P>1)OR(wfp_P =1) THEN(0)ELSE IF(wfp_P>wfp_fc_P) AND(wfp_P<1)THEN ((1-wfp_P)/(1-wfp_fc_P)) ELSE(1)	-	
mode_S	controlling factor for oxygen availability in S wetland	IF(wfp_S>1) OR(wfp_S=1) THEN(0)ELSE IF(wfp_S>wfp_fc_S) AND(wfp_S<1)THEN ((1-wfp_S)/(1-wfp_fc_S)) ELSE(1)	-	
NH4S_load	ammonium entering the S wetland	0.15	g P/day	estimate
NO3S_load	nitrate entering the S wetland	0.1	g P/day	estimate
NP_conc_in_plant	N concentration of plant in P wetland	(total_NP_in_plant/total_CP_in_plant)*C_conc_in_plant	g N/g DW	
NS_conc_in_plant	N concentration of plant in S wetland	(total_NS_in_plant/total_CS_in_plant)*C_conc_in_plant	g N/g DW	
N_conc_minimum	minimum N concentration required for growth	0.0016	g N/g DW	van der Peijl and Verhoeven, 1999
N_conc_optimal	optimal N concentration for plant growth	N_max_papyrus/ max_papyrus_biomass	g N/g DW	
N_max_AGB	maximum amount of nitrogen in aboveground biomass	max_papyrus_biomass* perc_N_in_AGB* ratio_AGB_total_biomass	g N/m^2	
N_max_BGB	maximum amount of nitrogen in belowground biomass	max_papyrus_biomass* perc_N_in_BGB* ratio_BGB_total_biomass	g N/m^2	

name	description	equation or value	unit	source
N_max_papyrus	maximum amount of nitrogen in papyrus	N_max_AGB+N_max_BGB	g N/m^2	
N_papyrus_P	total nitrogen in papyrus in P wetland	NAGBP+NBGBP	g N/m^2	
N_papyrus_S	total nitrogen in papyrus in S wetland	NAGBS+NBGBS	g N/m^2	
N_retrans_constant	fraction of nitrogen retranslocated after dying shoot	0.4	-	van der Peijl and Verhoeven, 1999
perc_C_in_AGB	initial mass percentage of carbon to dry weight of aboveground biomass	initial_C_AGB/initial_AGB	%	
perc_C_in_BGB	initial mass percentage of carbon to dry weight of belowground biomass	initial_C_BGB/initial_BGB	%	
perc_N_in_AGB	initial mass percentage of nitrogen to dry weight of aboveground biomass	initial_N_AGB/initial_AGB	%	
perc_N_in_BGB	initial mass percentage of nitrogen to dry weight of belowground biomass	initial_N_BGB/initial_BGB	%	
POC_hydrolysis_constant	hydrolysis rate for particulate organic carbon	$1*10^{-4}$	day^{-1}	estimate
POC_settling_rate	settling rate for particulate organic carbon	0.05	day-1	van Dam et al., 2007
PONS_load	particulate organic nitrogen entering the S wetland	0.1	g N/day	estimate

name	description	equation or value	unit	source
PON_hydrolysis_constant	hydrolysis rate for particulate organic nitrogen	$1*10^{-4}$	day^{-1}	estimate
PON_settling_constant	settling rate for particulate organic nitrogen	0.05	day-1	van Dam et al., 2007
porew_free_P	pore water minus pore water at field capacity in P wetland	IF(vol_frac_P>porosity_P)THEN ((1-wfp_fc_P)*porosity_P* soil_volume_P)ELSE IF(vol_frac_P>wfp_fc_P* porosity_P)AND(vol_frac_P<porosity_P)OR(vol_frac_P=porosity_P) THEN((vol_frac_P-wfp_fc_P*porosity_P) *soil_volume_P)ELSE(0)	m^3/m^2	
porew_free_S	pore water minus pore water at field capacity in S wetland	IF(vol_frac_S>porosity_S)THEN ((1-wfp_fc_S)*porosity_S* soil_volume_S)ELSE IF(vol_frac_S>wfp_fc_S* porosity_S)AND(vol_frac_S<porosity_S)OR(vol_frac_S=porosity_S) THEN((vol_frac_S-wfp_fc_S*porosity_S) *soil_volume_S)ELSE(0)	m^3/m^2	
pore_water_P	pore water in P wetland	WaterP-surfw_P	m^3/m^2	
pore_water_S	pore water in S wetland	WaterS-surfw_S	m^3/m^2	
porosity_P	porosity of the soil in P wetland	0.8	-	estimate
porosity_S	porosity of the soil in S wetland	0.8	-	estimate
radiance	mean monthly values for Naivasha	Counter (0, 365)	MJ/m^2* day	Muthuri et al., 1989

name	description	equation or value	unit	source
rainfall_rate	mean monthly values for Naivasha	Counter (0, 365)	mm/day	Gaudet, 1978
ratio_AGB_total_biomass	mass percentage of aboveground biomass of total papyrus	initial_AGB/ initial_papyrus_biomass	%	
ratio_BGB_total_biomass	mass percentage of belowground biomass of total papyrus	initial_BGB/ initial_papyrus_biomass	%	
river_inflow_rate	monthly means	Counter (0, 365)	m^3/day	estimate
soil_depth_P	soil depth, rooting depth in P wetland	0.2	m	estimate
soil_depth_S	soil depth, rooting depth in S wetland	0.2.	m	estimate
soil_volume_P	soil volume in P wetland	soil_depth_P*1	m^3/m^2	
soil_volume_S	soil volume in S wetland	soil_depth_S*1	m^3/m^2	
surfw_P	surface water in P wetland	IF(vol_frac_P<porosity_P) OR(vol_frac_P=porosity_P) THEN(0) ELSE((vol_frac_P-porosity_P)*soil_volume_P)	m^3/m^2	
surfw_S	surface water in S wetland	IF(vol_frac_S<porosity_S) OR(vol_frac_S=porosity_S) THEN(0) ELSE((vol_frac_S-porosity_S)*soil_volume_S)	m^3/m^2	
treshold_swd_P	maximum water depth of P wetland	0.5	m	estimate
total_CP_in_plant	total C in plant in P wetland	CAGBP+CBGBP	g C/m^2	
total_CS_in_plant	total C in plant in S wetland	CAGBS+CBGBS	g C/m^2	
total_NP_in_plant	total N in plant in P wetland	NAGBP+NBGBP	g N/m^2	
total_NS_in_plant	total N in plant in S wetland	NAGBS+NBGBS	g N/m^2	

name	description	equation or value	unit	source
vol_frac_P	volume fraction of pore water filled in P wetland	WaterP/soil_volume_P	-	
vol_frac_S	volume fraction of pore water filled in S wetland	WaterS/soil_volume_S	-	
wfp_fc_P	water filled porosity at field capacity in P wetland	0.625	-	estimate
wfp_fc_S	water filled porosity at field capacity in S wetland	0.625	-	estimate
wfp_P	water filled porosity in P wetland	vol_frac_P/porosity_P	-	
wfp_S	water filled porosity in S wetland	vol_frac_S/porosity_S	-	

Appendix 4.2: Model equations

hydrology
WaterP(t) = WaterP(t - dt) + (prec_P + surface_water_flow + lake_Inflow - evap_P - recharge_P - outflow) *
dt
INIT WaterP = 0.5
INFLOWS:
prec_P = rainfall_rate*0.001
surface_water_flow = froff_S*surfw_S
lake_Inflow = IF(surfw_P<threshold_swd_P) THEN(lake_inflow_rate) ELSE(0)
OUTFLOWS:
evap_P = evaporation_rate*0.001
recharge_P = frout_P*porew_free_P
outflow = IF(surfw_P>threshold_swd_P) THEN(surface_water_flow+prec_P-evap_P-recharge_P) ELSE(0)
WaterS(t) = WaterS(t - dt) + (prec_S + inflow_S - evap_S - recharge_S - surface_water_flow) * dt
INIT WaterS = 0.2
INFLOWS:
prec_S = rainfall_rate*0.001
inflow_S = river_inflow_rate/area_S
OUTFLOWS:
evap_S = evaporation_rate*0.001
recharge_S = frout_S*porew_free_S
surface_water_flow = froff_S*surfw_S
area_S = 1000
froff_S = 0.5
frout_P = 0.01
frout_S = 0.01
lake_inflow_rate = 0.12
mode_P = IF(wfp_P>1) OR(wfp_P=1) THEN(0) ELSE
IF(wfp_P>wfp_fc_P) AND(wfp_P<1) THEN((1-wfp_P)/(1-wfp_fc_P)) ELSE(1)
mode_S = IF(wfp_S>1) OR(wfp_S=1) THEN(0) ELSE
IF(wfp_S>wfp_fc_S) AND(wfp_S<1) THEN((1-wfp_S)/(1-wfp_fc_S)) ELSE(1)
porew_free_P = IF(vol_frac_P>porosity_P) THEN((1-wfp_fc_P)*porosity_P*soil_volume_P) ELSE
IF(vol_frac_P>wfp_fc_P*porosity_P) AND(vol_frac_P<porosity_P) OR(vol_frac_P=porosity_P)
THEN((vol_frac_P-wfp_fc_P*porosity_P)*soil_volume_P)
ELSE(0)
porew_free_S = IF(vol_frac_S>porosity_S) THEN((1-wfp_fc_S)*porosity_S*soil_volume_S) ELSE
IF(vol_frac_S>wfp_fc_S*porosity_S) AND(vol_frac_S<porosity_S) OR(vol_frac_S=porosity_S)
THEN((vol_frac_S-wfp_fc_S*porosity_S)*soil_volume_S)
ELSE(0)
pore_water_P = WaterP-surfw_P
pore_water_S = WaterS-surfw_S
porosity_P = 0.8
porosity_S = 0.8
soil_depth_P = 0.2
soil_depth_S = 0.2
soil_volume_P = soil_depth_P*1
soil_volume_S = soil_depth_S*1
surfw_P = IF(vol_frac_P<porosity_P) OR(vol_frac_P=porosity_P) THEN(0)

ELSE((vol_frac_P-porosity_P)*soil_volume_P)

surfw_S = IF(vol_frac_S<porosity_S) OR(vol_frac_S=porosity_S) THEN(0)

ELSE((vol_frac_S-porosity_S)*soil_volume_S)

threshold_swd_P = 0.5

vol_frac_P = WaterP/soil_volume_P

vol_frac_S = WaterS/soil_volume_S

wfp_fc_P = 0.625

wfp_fc_S = 0.625

wfp_P = vol_frac_P/porosity_P

wfp_S = vol_frac_S/porosity_S

evaporation_rate = GRAPH(COUNTER(0,365))

(1.00, 128), (31.3, 129), (61.7, 128), (92.0, 135), (122, 116), (153, 150), (183, 125), (213, 100), (244, 125),

(274, 137), (304, 144), (335, 105), (365, 131)

rainfall_rate = GRAPH(Counter (0, 365))

(0.00, 3.60), (30.4, 4.50), (60.8, 10.0), (91.3, 60.0), (122, 190), (152, 90.0), (183, 75.0), (213, 75.0), (243,

40.0), (274, 40.0), (304, 25.0), (335, 40.0), (365, 25.0)

river_inflow_rate = GRAPH(Counter (0, 365))

(0.00, 112), (30.4, 135), (60.8, 151), (91.3, 156), (122, 193), (152, 162), (183, 135), (213, 135), (243, 85.8),

(274, 96.3), (304, 112), (335, 89.3), (365, 91.0)

papyrus biomass

initial_AGB = 3489

initial_BGB = 3928

initial_C_AGB = 1853

initial_C_BGB = 1570

initial_C_DAGB = initial_dead_papyrus_biomass*ratio_AGB_total_biomass*perc_C_in_AGB

initial_C_DBGB = initial_dead_papyrus_biomass*ratio_BGB_total_biomass*perc_C_in_BGB

initial_dead_papyrus_biomass = 1340

initial_N_AGB = 44

initial_N_BGB = 31

initial_N_DAGB = initial_dead_papyrus_biomass*ratio_AGB_total_biomass*perc_N_in_AGB

initial_N_DBGB = initial_dead_papyrus_biomass*ratio_BGB_total_biomass*perc_N_in_BGB

initial_papyrus_biomass = initial_AGB+initial_BGB

max_papyrus_biomass = 8118

N_max_AGB = max_papyrus_biomass*perc_N_in_AGB*ratio_AGB_total_biomass

N_max_BGB = max_papyrus_biomass*perc_N_in_BGB*ratio_BGB_total_biomass

N_max_papyrus = N_max_AGB+N_max_BGB

perc_C_in_AGB = initial_C_AGB/initial_AGB

perc_C_in_BGB = initial_C_BGB/initial_BGB

perc_N_in_AGB = initial_N_AGB/initial_AGB

perc_N_in_BGB = initial_N_BGB/initial_BGB

ratio_AGB_total_biomass = initial_AGB/initial_papyrus_biomass

ratio_BGB_total_biomass = initial_BGB/initial_papyrus_biomass

permanently flooded nitrogen

DONPP(t) = DONPP(t - dt) + (PONPP_hydrolysis + DONP_diffusion - DONPP_mineral -

DONP_recharge)

* dt

INIT DONPP = 1.1
INFLOWS:
PONPP_hydrolysis = PONPP*PON_hydrolysis_constant
DONP_diffusion = IF(surfw_P>0)
THEN(K_DON_diffusion*((DONSP/surfw_P)-(DONPP/soil_depth_P))/((surfw_P+soil_depth_P)/2))
ELSE(0)
OUTFLOWS:
DONPP_mineral = K_mineral*DONPP
DONP_recharge = conc_DONPP*recharge_P
DONSP(t) = DONSP(t - dt) + (PONSP_hydrolysis + DON_w_surf_in + DON_w_lake_in -
DONSP_mineral -
DONP_diffusion - DONSP_outflow) * dt
INIT DONSP = 1.1
INFLOWS:
PONSP_hydrolysis = PONSP*PON_hydrolysis_constant
DON_w_surf_in = surface_water_flow*conc_DONS
DON_w_lake_in = lake_Inflow*conc_DON__lake_inflow
OUTFLOWS:
DONSP_mineral = K_mineral*DONSP
DONP_diffusion = IF(surfw_P>0)
THEN(K_DON_diffusion*((DONSP/surfw_P)-(DONPP/soil_depth_P))/((surfw_P+soil_depth_P)/2))
ELSE(0)
DONSP_outflow = outflow*conc_DONSP
NAGBP(t) = NAGBP(t - dt) + (- N_trans_P - N_retrans_P - NAGBP_death - NAGBP_harvest_D) * dt
INIT NAGBP = initial_N_AGB
OUTFLOWS:
N_trans_P = (NAGBP-NBGBP*C_AGB_to_BGB_ratio_P)/(C_AGB_to_BGB_ratio_P+1)
N_retrans_P = CAGBP_death*N_retrans_constant/CN_AGBP_ratio
NAGBP_death = CAGBP_death*(1-N_retrans_constant)/CN_AGBP_ratio
NAGBP_harvest_D = CAGBP_harvest_D/CN_AGBP_ratio
NBGBP(t) = NBGBP(t - dt) + (N_trans_P + N_retrans_P + NH4_uptake_P + NO3_uptake_P -
NBGBP_death) * dt
INIT NBGBP = initial_N_BGB
INFLOWS:
N_trans_P = (NAGBP-NBGBP*C_AGB_to_BGB_ratio_P)/(C_AGB_to_BGB_ratio_P+1)
N_retrans_P = CAGBP_death*N_retrans_constant/CN_AGBP_ratio
NH4_uptake_P =
max_NH4_uptake*N_papyrus_P*(1-N_papyrus_P/N_max_papyrus)*limit_NH4PP
NO3_uptake_P =
max_NO3_uptake*N_papyrus_P*(1-N_papyrus_P/N_max_papyrus)*limit_NO3PP
OUTFLOWS:
NBGBP_death = CBGBP_death/CN_BGBP_ratio
NDAGBP(t) = NDAGBP(t - dt) + (NAGBP_death - NDAGBP_frag - NDAGBP_leach) * dt
INIT NDAGBP = initial_N_DAGB
INFLOWS:
NAGBP_death = CAGBP_death*(1-N_retrans_constant)/CN_AGBP_ratio
OUTFLOWS:
NDAGBP_frag = CDAGBP_frag/CN_DAGBP_ratio
NDAGBP_leach = CDAGBP_leach/CN_DAGBP_ratio
NDBGBP(t) = NDBGBP(t - dt) + (NBGBP_death - NDBGBP_frag - NDBGBP_leach) * dt
INIT NDBGBP = initial_N_DBGB

INFLOWS:

NBGBP_death = CBGBP_death/CN_BGBP_ratio

OUTFLOWS:

NDBGBP_frag = CDBGBP_frag/CN_DBGBP_ratio

NDBGBP_leach = CDBGBP_leach/CN_DBGBP_ratio

NH4AP(t) = NH4AP(t - dt) + (NH4P_adsorption) * dt

INIT NH4AP = 0.5

INFLOWS:

NH4P_adsorption =

(IF(conc_NH4PP>5)THEN(NH4PP*K_NH4_adsorption)ELSE(-2*K_NH4_adsorption*NH4AP))

NH4PP(t) = NH4PP(t - dt) + (DONPP_mineral + NH4P_diffusion + NDBGBP_leach - nitrifi_PP -

NH4_uptake_P - NH4P_recharge - NH4P_adsorption) * dt

INIT NH4PP = 0.5

INFLOWS:

DONPP_mineral = K_mineral*DONPP

NH4P_diffusion = IF(surfw_P>0)

THEN(K_NH4_diffusion*((NH4SP/surfw_P)-(NH4PP/soil_depth_P))/((surfw_P+soil_depth_P)/2))

ELSE(0)

NDBGBP_leach = CDBGBP_leach/CN_DBGBP_ratio

OUTFLOWS:

nitrifi_PP = K_nitri*mode_P*NH4PP

NH4_uptake_P =

max_NH4_uptake*N_papyrus_P*(1-N_papyrus_P/N_max_papyrus)*limit_NH4PP

NH4P_recharge = conc_NH4PP*recharge_P

NH4P_adsorption =

(IF(conc_NH4PP>5)THEN(NH4PP*K_NH4_adsorption)ELSE(-2*K_NH4_adsorption*NH4AP))

NH4SP(t) = NH4SP(t - dt) + (DONSP_mineral + NH4S_in + NH4_lake_in + NDAGBP_leach - nitri_SP -

NH4P_diffusion - NH4SP_outflow) * dt

INIT NH4SP = 0.5

INFLOWS:

DONSP_mineral = K_mineral*DONSP

NH4S_in = surface_water_flow*conc_NH4S

NH4_lake_in = lake_Inflow*conc_NH4_lake_inflow

NDAGBP_leach = CDAGBP_leach/CN_DAGBP_ratio

OUTFLOWS:

nitri_SP = K_nitri*NH4SP

NH4P_diffusion = IF(surfw_P>0)

THEN(K_NH4_diffusion*((NH4SP/surfw_P)-(NH4PP/soil_depth_P))/((surfw_P+soil_depth_P)/2))

ELSE(0)

NH4SP_outflow = outflow*conc_NH4SP

NO3PP(t) = NO3PP(t - dt) + (nitrifi_PP + NO3P_diffusion - NO3_uptake_P - NO3P_recharge -

denitri_PP)

dt

INIT NO3PP = 0.05

INFLOWS:

nitrifi_PP = K_nitri*mode_P*NH4PP

NO3P_diffusion = IF(surfw_P>0)

THEN(K_NO3_diffusion*((NO3SP/surfw_P)-(NO3PP/soil_depth_P))/((surfw_P+soil_depth_P)/2))

ELSE(0)

OUTFLOWS:

NO3_uptake_P =

max_NO3_uptake*N_papyrus_P*(1-N_papyrus_P/N_max_papyrus)*limit_NO3PP

NO3P_recharge = conc_NO3PP*recharge_P

denitri_PP = K_denitri*NO3PP*(1-mode_P)

NO3SP(t) = NO3SP(t - dt) + (nitri_SP + NO3S_in + NO3__lake_in - NO3P_diffusion - NO3SP_outflow) * dt

INIT NO3SP = 0.05

INFLOWS:

nitri_SP = K_nitri*NH4SP

NO3S_in = surface_water_flow*conc_NO3S

NO3__lake_in = lake_Inflow*conc_NO3__lake_inflow

OUTFLOWS:

NO3P_diffusion = IF(surfw_P>0)

THEN(K_NO3_diffusion*((NO3SP/surfw_P)-(NO3PP/soil_depth_P))/((surfw_P+soil_depth_P)/2))

ELSE(0)

NO3SP_outflow = outflow*conc_NO3SP

PONPP(t) = PONPP(t - dt) + (NDBGBP_frag + PONSP_settling - PONPP_hydrolysis - PONP_recharge) * dt

INIT PONPP = 0.9

INFLOWS:

NDBGBP_frag = CDBGBP_frag/CN_DBGBP_ratio

PONSP_settling = PONSP*PON_settling_constant

OUTFLOWS:

PONPP_hydrolysis = PONPP*PON_hydrolysis_constant

PONP_recharge = conc_PONPP*recharge_P

PONSP(t) = PONSP(t - dt) + (NDAGBP_frag + PONSP__surf_in + PONSP_lake_in - PONSP_hydrolysis - PONSP_settling - PONSP_outflow) * dt

INIT PONSP = 0.9

INFLOWS:

NDAGBP_frag = CDAGBP_frag/CN_DAGBP_ratio

PONSP__surf_in = surface_water_flow*conc_PONS

PONSP_lake_in = lake_Inflow*conc_PON__lake_inflow

OUTFLOWS:

PONSP_hydrolysis = PONSP*PON_hydrolysis_constant

PONSP_settling = PONSP*PON_settling_constant

PONSP_outflow = outflow*conc_PONSP

CN_AGBP_ratio = CAGBP/NAGBP

CN_BGBP_ratio = CBGBP/NBGBP

CN_DAGBP_ratio = CDAGBP/NDAGBP

CN_DBGBP_ratio = CDBGBP/NDBGBP

conc_DONPP = DONPP/pore_water_P

conc_DONSP = DONSP/surfw_P

conc_DON__lake_inflow = 0.065

conc_NH4PP = NH4PP/pore_water_P

conc_NH4SP = NH4SP/surfw_P

conc_NH4_lake_inflow = 0.8

conc_NO3PP = NO3PP/pore_water_P

conc_NO3SP = NO3SP/surfw_P

conc_NO3__lake_inflow = 0.5

conc_PONPP = PONPP/pore_water_P

conc_PONSP = PONSP/surfw_P

conc_PON__lake_inflow = 0.065

K_DON_diffusion = 0.05
K_NH4_diffusion = 0.05
K_NO3_diffusion = 0.05
limit_NH4PP = conc_NH4PP/(conc_NH4PP+K_NH4)
limit_NO3PP = conc_NO3PP/(conc_NO3PP+K_NO3)
N_papyrus_P = NAGBP+NBGBP
PON_settling_constant = 0.05

permanently flooded carbon
CAGBP(t) = CAGBP(t - dt) + (CAGBP__assimilation - C_trans_P - CAGBP_death - CAGBP_respiration - CAGBP_harvest_D) * dt
INIT CAGBP = initial_C_AGB
INFLOWS:
CAGBP__assimilation =
max_assimilation_constant*CAGBP*limit_radiance*(limit_N_P/0.9)*((max_AGB_biomass*perc_C_in_AGB-CAGBP)/(max_AGB_biomass*perc_C_in_AGB))
OUTFLOWS:
C_trans_P = (CAGBP-CBGBP*C_AGB_to_BGB_optimal_ratio)/(1+C_AGB_to_BGB_optimal_ratio)
CAGBP_death = CAGBP*CAGB_death_constant
CAGBP_respiration = CAGBP*maint_coeff_AGB+CAGBP__assimilation*growth_coeff_AGB
CAGBP_harvest_D = PULSE(harvest_in_g_C, harvest_day-1,harvest_interval)
CBGBP(t) = CBGBP(t - dt) + (C_trans_P - CBGBP_death - CBGBP_respiration) * dt
INIT CBGBP = initial_C_BGB
INFLOWS:
C_trans_P = (CAGBP-CBGBP*C_AGB_to_BGB_optimal_ratio)/(1+C_AGB_to_BGB_optimal_ratio)
OUTFLOWS:
CBGBP_death = CBGBP*CBGB_death_constant
CBGBP_respiration = CBGBP*maint_coeff_BGB+C_trans_P*growth_coeff_BGB
CDAGBP(t) = CDAGBP(t - dt) + (CAGBP_death - CDAGBP_frag - CDAGBP_leach) * dt
INIT CDAGBP = initial_C_DAGB
INFLOWS:
CAGBP_death = CAGBP*CAGB_death_constant
OUTFLOWS:
CDAGBP_frag = CDAGBP*CDAGB_frag_constant
CDAGBP_leach = CDAGB_leach_constant*CAGBP_death
CDBGBP(t) = CDBGBP(t - dt) + (CBGBP_death - CDBGBP_frag - CDBGBP_leach) * dt
INIT CDBGBP = initial_C_DBGB
INFLOWS:
CBGBP_death = CBGBP*CBGB_death_constant
OUTFLOWS:
CDBGBP_frag = CDBGBP*CDBGB_frag_constant
CDBGBP_leach = CDBGB_leach_constant*CBGBP_death
POCPP(t) = POCPP(t - dt) + (CDBGBP_frag + POCSP_settling - POCPP_hydrolysis) * dt
INIT POCPP = 20
INFLOWS:
CDBGBP_frag = CDBGBP*CDBGB_frag_constant
POCSP_settling = POCSP*POC_settling_rate
OUTFLOWS:
POCPP_hydrolysis = POCPP*POC_hydrolysis_constant
POCSP(t) = POCSP(t - dt) + (CDAGBP_frag - POCSP_hydrolysis - POCSP_settling) * dt
INIT POCSP = 20

INFLOWS:
CDAGBP_frag = CDAGBP*CDAGB_frag_constant
OUTFLOWS:
POCSP_hydrolysis = POCSP*POC_hydrolysis_constant
POCSP_settling = POCSP*POC_settling_rate
AGBP = CAGBP/perc_C_in_AGB
BGBP = CBGBP/perc_C_in_BGB
C_AGB_to_BGB_ratio_P = CAGBP/CBGBP
limit_N_P =
IF((NP_conc_in_plant-N_conc_minimum)/((kmN-N_conc_minimum)+(NP_conc_in_plant-
N_conc_minimu
m))<(0.9-NP_conc_in_plant))AND(0.9-NP_conc_in_plant)>N_conc_minimum
THEN(NP_conc_in_plant-N_conc_minimum)/((kmN-N_conc_minimum)+(NP_conc_in_plant-
N_conc_mini
mum))ELSE
IF((NP_conc_in_plant-N_conc_minimum)/((kmN-N_conc_minimum)+(NP_conc_in_plant-
N_conc_minimu
m))>=(0.9-NP_conc_in_plant))AND((0.9-NP_conc_in_plant)>N_conc_minimum)
THEN(0.9)ELSE(0)
NP_conc_in_plant = (total_NP_in_plant/total_CP_in_plant)*C_conc_in_plant
POC_settling_rate = 0.05
total_CP_in_plant = CAGBP+CBGBP
total_NP_in_plant = NAGBP+NBGBP

seasonally flooded carbon
CAGBS(t) = CAGBS(t - dt) + (CAGBS_assimilation - CAGBS_death - C_trans_S - CAGBS_respiration -
CAGBS_harvest_D) * dt
INIT CAGBS = initial_C_AGB
INFLOWS:
CAGBS_assimilation =
max_assimilation_constant*CAGBS*(limit_N_S/0.9)*limit_radiance*((max_AGB_biomass*perc_C_i
n_AGB-CAGBS)/(max_AGB_biomass*perc_C_in_AGB))
OUTFLOWS:
CAGBS_death = (CAGBS*CAGB_death_constant)
C_trans_S =
((CAGBS-CBGBS*C_AGB_to_BGB_optimal_ratio)/(1+C_AGB_to_BGB_optimal_ratio))
CAGBS_respiration = (CAGBS*maint_coeff_AGB+CAGBS_assimilation*growth_coeff_AGB)
CAGBS_harvest_D = (PULSE(harvest_in_g_C,
harvest_day-1,harvest_interval))*harvest_in_S_yes_or_no
CBGBS(t) = CBGBS(t - dt) + (C_trans_S - CBGBS_death - CBGBS_respiration) * dt
INIT CBGBS = initial_C_BGB
INFLOWS:
C_trans_S =
((CAGBS-CBGBS*C_AGB_to_BGB_optimal_ratio)/(1+C_AGB_to_BGB_optimal_ratio))
OUTFLOWS:
CBGBS_death = CBGBS*CBGB_death_constant
CBGBS_respiration = CBGBS*maint_coeff_BGB+C_trans_S*growth_coeff_BGB
CDAGBS(t) = CDAGBS(t - dt) + (CAGBS_death - CDAGBS_frag - CDAGBS_leach) * dt
INIT CDAGBS = initial_C_DAGB
INFLOWS:
CAGBS_death = (CAGBS*CAGB_death_constant)

OUTFLOWS:

CDAGBS_frag = CDAGBS*CDAGB_frag_constant

CDAGBS_leach = CDAGB_leach_constant*CAGBS_death

CDBGBS(t) = CDBGBS(t - dt) + (CBGBS_death - CDBGBS_frag - CDBGBS_leach) * dt

INIT CDBGBS = initial_C_DBGB

INFLOWS:

CBGBS_death = CBGBS*CBGB_death_constant

OUTFLOWS:

CDBGBS_frag = CDBGBS*CDBGB_frag_constant

CDBGBS_leach = CDBGB_leach_constant*CBGBS_death

POCS(t) = POCS(t - dt) + (CDAGBS_frag + CDBGBS_frag - POCS___hydrolysis) * dt

INIT POCS = 10000

INFLOWS:

CDAGBS_frag = CDAGBS*CDAGB_frag_constant

CDBGBS_frag = CDBGBS*CDBGB_frag_constant

OUTFLOWS:

POCS___hydrolysis = POCS*POC_hydrolysis_constant

AGBS = CAGBS/perc_C_in_AGB

BGBS = CBGBS/perc_C_in_BGB

CAGB_death_constant = 0.04/7

CBGB_death_constant = 0.01/7

CDAGB_frag_constant = 0.00193/7

CDAGB_leach_constant = 0.432

CDBGB_frag_constant = 0.00584/7

CDBGB_leach_constant = 0.486

C_AGB_to_BGB_optimal_ratio = initial_C_AGB/initial_C_BGB

C_AGB_to_BGB_ratio_S = CAGBS/CBGBS

C_conc_in_plant = (initial_C_AGB+initial_C_BGB)/(initial_AGB+initial_BGB)

growth_coeff_AGB = 0.3

growth_coeff_BGB = 0.2

harvest_% = 0.1

harvest_day = 1

harvest_interval = 1

harvest_in_g_C = harvest_in__g_papyrus*perc_C_in_AGB

harvest_in_S_yes_or_no = 0

harvest_in__g_papyrus = 0

kmN = N_conc_minimum+((N_conc_optimal-N_conc_minimum)/9)

K_radiance = 1

limit_N_S =

IF((NS_conc_in_plant-N_conc_minimum)/((kmN-N_conc_minimum)+(NS_conc_in_plant-N_conc_minimu

m))<(0.9-NS_conc_in_plant))AND(0.9-NS_conc_in_plant)>N_conc_minimum

THEN(NS_conc_in_plant-N_conc_minimum)/((kmN-N_conc_minimum)+(NS_conc_in_plant-N_conc_mini

mum))ELSE

IF((NS_conc_in_plant-N_conc_minimum)/((kmN-N_conc_minimum)+(NS_conc_in_plant-N_conc_minimu

m))>=(0.9-NS_conc_in_plant))AND((0.9-NS_conc_in_plant)>N_conc_minimum)

THEN(0.9)ELSE(0)

limit_radiance = radiance/(radiance+K_radiance)

maint_coeff_AGB = 0.02

maint_coeff_BGB = 0.002
max_AGB_biomass = max_papyrus_biomass*ratio_AGB_total_biomass
max_assimilation_constant = 0.5
NS_conc_in_plant = (total_NS_in_plant/total_CS_in_plant)*C_conc_in_plant
N_conc_minimum = 0.0016
N_conc_optimal = N_max_papyrus/max_papyrus_biomass
POC_hydrolysis_constant = 0.0001
total_CS_in_plant = CAGBS+CBGBS
total_NS_in_plant = NAGBS+NBGBS
radiance = GRAPH(Counter (0, 365))
(0.00, 23.5), (30.4, 23.5), (60.8, 23.0), (91.3, 23.5), (122, 21.0), (152, 20.5), (183, 20.5), (213, 19.0), (243,
20.5), (274, 22.0), (304, 22.0), (335, 21.0), (365, 22.5)

seasonally flooded nitrogen
DONS(t) = DONS(t - dt) + (PONS_hydrolysis + DONS_inflow - DONS_mineral - DONS_recharge -
DONS_outflow) * dt
INIT DONS = 1.1
INFLOWS:
PONS_hydrolysis = PONS*PON_hydrolysis_constant
DONS_inflow = DONS_load
OUTFLOWS:
DONS_mineral = DONS*K_mineral
DONS_recharge = conc_DONS*recharge_S
DONS_outflow = surface_water_flow*conc_DONS
NAGBS(t) = NAGBS(t - dt) + (- NAGBS_death - N_retrans_S - N_trans_S - NAGBS__harvest_D) * dt
INIT NAGBS = initial_N_AGB
OUTFLOWS:
NAGBS_death = CAGBS_death*(1-N_retrans_constant)/CN_AGBS_ratio
N_retrans_S = CAGBS_death*N_retrans_constant/CN_AGBS_ratio
N_trans_S = (NAGBS-NBGBS*C_AGB_to_BGB_ratio_S)/(C_AGB_to_BGB_ratio_S+1)
NAGBS__harvest_D = CAGBS_harvest_D/CN_AGBS_ratio
NBGBS(t) = NBGBS(t - dt) + (N_retrans_S + NH4S_uptake + NO3S_uptake + N_trans_S -
NBGBS_death)
* dt
INIT NBGBS = initial_N_BGB
INFLOWS:
N_retrans_S = CAGBS_death*N_retrans_constant/CN_AGBS_ratio
NH4S_uptake = max_NH4_uptake*N_papyrus_S*(1-N_papyrus_S/N_max_papyrus)*limit_NH4S
NO3S_uptake = max_NO3_uptake*N_papyrus_S*(1-N_papyrus_S/N_max_papyrus)*limit_NO3S
N_trans_S = (NAGBS-NBGBS*C_AGB_to_BGB_ratio_S)/(C_AGB_to_BGB_ratio_S+1)
OUTFLOWS:
NBGBS_death = CBGBS_death/CN_BGBS_ratio
NDAGBS(t) = NDAGBS(t - dt) + (NAGBS_death - NDAGBS_frag - NDAGBS_leach) * dt
INIT NDAGBS = initial_N_DAGB
INFLOWS:
NAGBS_death = CAGBS_death*(1-N_retrans_constant)/CN_AGBS_ratio
OUTFLOWS:
NDAGBS_frag = CDAGBS_frag/CN_DAGBS_ratio
NDAGBS_leach = CDAGBS_leach/CN_DAGBS_ratio
NDBGBS(t) = NDBGBS(t - dt) + (NBGBS_death - NDBGBS_frag - NDBGBS_leach) * dt

INIT NDBGBS = initial_N_DBGB
INFLOWS:
NBGBS_death = CBGBS_death/CN_BGBS_ratio
OUTFLOWS:
NDBGBS_frag = CDBGBS_frag/CN_DBGBS_ratio
NDBGBS_leach = CDBGBS_leach/CN_DBGBS_ratio
NH4AS(t) = NH4AS(t - dt) + (NH4S_adsorption) * dt
INIT NH4AS = 5
INFLOWS:
NH4S_adsorption = (IF(conc_NH4S>5) THEN(K_NH4_adsorption*NH4S)
ELSE(-K_NH4_adsorption*NH4AS))
NH4S(t) = NH4S(t - dt) + (DONS_mineral + NDAGBS_leach + NDBGBS_leach + NH4S_inflow - nitri_S -
NH4S_uptake - NH4S_outflow - NH4S_recharge - NH4S_adsorption) * dt
INIT NH4S = 0.5
INFLOWS:
DONS_mineral = DONS*K_mineral
NDAGBS_leach = CDAGBS_leach/CN_DAGBS_ratio
NDBGBS_leach = CDBGBS_leach/CN_DBGBS_ratio
NH4S_inflow = NH4S_load
OUTFLOWS:
nitri_S = K_nitri*mode_S*NH4S
NH4S_uptake = max_NH4_uptake*N_papyrus_S*(1-N_papyrus_S/N_max_papyrus)*limit_NH4S
NH4S_outflow = surface_water_flow*conc_NH4S
NH4S_recharge = conc_NH4S*recharge_S
NH4S_adsorption = (IF(conc_NH4S>5) THEN(K_NH4_adsorption*NH4S)
ELSE(-K_NH4_adsorption*NH4AS))
NO3S(t) = NO3S(t - dt) + (nitri_S + NO3S_inflow - NO3S_uptake - denitri_S - NO3S_outflow -
NO3S_recharge) * dt
INIT NO3S = 0.05
INFLOWS:
nitri_S = K_nitri*mode_S*NH4S
NO3S_inflow = NO3S_load
OUTFLOWS:
NO3S_uptake = max_NO3_uptake*N_papyrus_S*(1-N_papyrus_S/N_max_papyrus)*limit_NO3S
denitri_S = K_denitri*NO3S*(1-mode_S)
NO3S_outflow = surface_water_flow*conc_NO3S
NO3S_recharge = conc_NO3S*recharge_S
PONS(t) = PONS(t - dt) + (NDAGBS_frag + NDBGBS_frag + PONS_inflow - PONS_hydrolysis -
PONS__outflow - PONS_recharge) * dt
INIT PONS = 0.9
INFLOWS:
NDAGBS_frag = CDAGBS_frag/CN_DAGBS_ratio
NDBGBS_frag = CDBGBS_frag/CN_DBGBS_ratio
PONS_inflow = PONS_load
OUTFLOWS:
PONS_hydrolysis = PONS*PON_hydrolysis_constant
PONS__outflow = surface_water_flow*conc_PONS
PONS_recharge = conc_PONS*recharge_S
CN_AGBS_ratio = CAGBS/NAGBS
CN_BGBS_ratio = CBGBS/NBGBS
CN_DAGBS_ratio = CDAGBS/NDAGBS

CN_DBGBS_ratio = CDBGBS/NDBGBS
conc_DONS = IF(WaterS>0) THEN(DONS/WaterS) ELSE(0)
conc_DONS_inflow = DONS_load/inflow_S
conc_NH4S = IF(WaterS>0) THEN(NH4S/WaterS) ELSE(0)
conc_NH4S_inflow = NH4S_load/inflow_S
conc_NO3S = IF(WaterS>0) THEN(NO3S/WaterS) ELSE(0)
conc_NO3S_inflow = NO3S_load/inflow_S
conc_PONS = IF(WaterS>0) THEN(PONS/WaterS) ELSE(0)
conc_PONS_inflow = PONS_load/inflow_S
DONS_load = 0.1
K_denitri = 0.01
K_mineral = 0.0002
K_NH4 = 0.7
K_NH4_adsorption = 0.01
K_nitri = 0.005
K_NO3 = 0.1
limit_NH4S = conc_NH4S/(conc_NH4S+K_NH4)
limit_NO3S = conc_NO3S/(conc_NO3S+K_NO3)
max_NH4_uptake = 0.05
max_NO3_uptake = 0.05
NH4S_load = 0.15
NO3S_load = 0.1
N_papyrus_S = NAGBS+NBGBS
N_retrans_constant = 0.4
PONS_load = 0.1
PON_hydrolysis_constant = 0.04

5

MODELLING NITROGEN AND PHOSPHORUS CYCLING AND RETENTION IN *CYPERUS PAPYRUS* DOMINATED NATURAL WETLANDS[4]

ABSTRACT

Papyrus (*Cyperus papyrus*) wetlands support millions through food provisioning, which leads to loss of regulating ecosystem services. This study aimed at understanding the impact of changes in water regime and vegetation harvesting on nutrient retention in rooted papyrus wetlands. A simulation model (Papyrus Simulator), developed and calibrated with data from African wetlands, produced reasonable estimates of productivity and nutrient retention. Phosphorus retention was lower than nitrogen retention, leading to a nitrogen-limited environment by reducing the N:P ratio in the water. Absence of surface water during part of the year caused a reduction of biomass. Harvesting increased nitrogen retention from 7% to over 40%, and phosphorus retention from 4% to 40%. Sensitivity analysis revealed assimilation, mortality, decay, re-translocation, nutrient inflow and soil porosity as the most influential factors. Papyrus Simulator is suitable for studying nutrient retention and harvesting in wetlands, and contributes to quantification of ecosystem services and sustainable wetland management.

Key words: papyrus; nitrogen retention; phosphorus retention; modelling; ecosystem services; wetlands

---

[4] Published as:

Hes EMA, van Dam AA (2019) Modelling nitrogen and phosphorus cycling and retention in *Cyperus papyrus* dominated natural wetlands. Environmental Modelling & Software 122:104531.

5.1 INTRODUCTION

In Africa, where so many people depend directly on wetlands for their livelihoods (Schuyt 2005; Maclean et al. 2014), population growth, climate change, the need for food security (Conceição et al. 2016), the suitability of wetlands for food production, and weak implementation of wetland conservation policies create an enormous pressure on wetlands (Rebelo et al. 2010, Davidson 2014). Current per capita food production in Africa is at the level of the 1960s (Pretty et al. 2011). More than one in four Africans are undernourished (UNDP 2012).

Pretty et al. (2011) defined sustainable agricultural intensification as 'producing more output from the same area of land while reducing negative environmental impacts and increasing contributions to natural capital and the value of environmental services', and identified nutrient cycling as one of its attributes. To apply this concept to agriculture in and around wetlands, more knowledge is needed about the dynamics of nutrient retention, both in relation to human activities (e.g. agriculture and vegetation harvesting, or extreme weather events due to climate change) and to natural variation in wetland processes (e.g. seasonal changes in hydrology). This knowledge can be used to develop more effective management practices to increase food security while protecting important ecosystem services.

The still widespread papyrus (*Cyperus papyrus*) wetlands in East Africa (van Dam et al. 2014; Ajwang' Ondiek et al. 2016) provide a wide range of ecosystem services. These wetlands support the livelihoods of millions of people through provisioning services like seasonal agriculture, papyrus harvesting, drinking water, clay and sand mining, fishing, and fuel production (Morrison et al. 2012; Morrison et al. 2014; Jones et al. 2016; van Dam and Kipkemboi 2018). The regulating services in these wetlands reduce the runoff of nutrients, sediments and trace elements into lakes and rivers (Kansiime et al. 2007) and regulate water quantity (flood protection and water storage) and local climate (e.g. by influencing rainfall patterns). Due to its C_4 photosynthesis, papyrus vegetation can sequester up to 0.48 kg C m^{-2} y^{-1}, storing up to 8.8 kg C m^{-2} in biomass and 64 kg C m^{-2} in detrital and peat deposits (Saunders et al. 2007; 2014). After clearing the aboveground biomass, a full stand of papyrus can grow back within 6-9 months (Muthuri et al. 1989, Terer et al. 2012a; Terer et al. 2012b). This high productivity provides opportunity for seasonal crop production in the dry season with recovery of the papyrus vegetation through rhizomes, and continuation of regulating services like nitrogen and phosphorus retention, in the wet season (van Dam et al. 2014).

The processes underlying nitrogen (N) and phosphorus (P) retention (accumulation of organic matter, uptake, nitrification, denitrification, adsorption) depend on conditions resulting from the hydrology and the presence of vegetation in the wetland, and their interaction (Powers et al. 2012). Permanently flooded areas favour peat formation and denitrification, while seasonally flooded zones favour nitrification and are more prone to anthropogenic disturbance. In papyrus between 45 and 105 g N m^{-2} is stored in living biomass, and 105 – 457 g N m^{-2} in dead biomass and peat (Gaudet 1977; Gaudet and Muthuri 1981; Boar et al. 1999;

Boar 2006). Potential denitrification in floating papyrus in Lake Naivasha ranged from 2.3 to 6.4 g N m^{-2} y^{-1} (Viner 1982). Reported values for P storage in living biomass were 5.4 g P m^{-2}, and in detritus and peat 3 – 57 g P m^{-2} (Gaudet 1977; Gaudet and Muthuri 1981; Boar 2006). For P adsorption, Kelderman et al. (2007) found a maximum adsorption of 4 mg P g^{-1} sediment in Kirinya wetland (Uganda). Lakes are often downstream of papyrus wetlands and Guildford and Hecky (2000) found Lake Victoria to be N limited and therefore favouring N-fixing cyanobacterial blooms at higher P concentrations. Lakes are more often N limited, but not exclusively and limitation can shift from N to P with increased global N deposition rates (Elser et al. 2009). It is, therefore, crucial to understand N and P-cycling in papyrus wetlands and how wetlands influence the N:P ratio in runoff water.

Simulation models can help to improve understanding of the complex dynamics of nutrient retention as influenced by natural (e.g. seasonal hydrology) and anthropogenic (e.g. agriculture and harvesting) drivers of change. Earlier model studies analysed the retention capacity of Lake Victoria fringing wetlands with a focus on processes for N, P and organic matter (Mwanuzi et al. 2003); and N processes for floating papyrus wetlands (van Dam et al. 2007). While some studies have concluded that papyrus wetlands are effective in removing or retaining N and P, e.g. up to 35% of total N from the inflowing water (Kanyiginya et al. 2010) and inorganic P up to 90% (Mwanuzi et al. 2003), the understanding of underlying processes is still limited. There is no evidence that these conclusions are valid over a longer period or merely a result of temporal N and P storage. Besides simulating changing conditions on a local scale (e.g. vegetation harvesting), process models can improve regional and global models to estimate the impact of human activities on biodiversity and ecosystem services (Sjögersten et al. 2014; Janse et al. 2015; Beusen et al. 2016, Costanza et al. 2017).

In a previous study, we constructed a simulation model for N cycling in natural rooted papyrus wetlands (Chapter 4). The model (now called Papyrus Simulator) showed that N retention increased, from 13% at low harvesting to a maximum of 50% per year at intermediate harvesting rates. If harvesting was increased further, N retention dropped dramatically (<5%) as harvesting exceeded regrowth. The model did not incorporate P and was not able to compare permanently and seasonally flooded zones independently. The overall objective of the current study is to understand the impact of changes in water regime and papyrus harvesting on N and P retention in rooted papyrus wetlands. The specific objectives are: (1) to further develop Papyrus Simulator by incorporating P processes and a hydrology section that enables independent comparison of different hydrological conditions and make it generally applicable to papyrus dominated wetlands; (2) to conduct a sensitivity analysis to identify which processes and parameters are most important in retaining N and P and to which parameters the model is most sensitive; and (3) to compare Papyrus Simulator outputs with published field data.

5.2 METHODS

5.2.1 MODEL DESCRIPTION AND DEVELOPMENT

The original Papyrus Simulator described a rooted papyrus wetland (Chapter 4) and modelled the influence of papyrus growth and harvesting under different hydrological conditions on the cycling of N (Figure 5.1). To achieve this, the model comprised three interacting sections: hydrology, carbon, and nitrogen (Figure 5.2). In the hydrology section, water level and soil moisture were calculated based on a simple water balance model. Soil moisture conditions determined the factor MODE (representing conditions from anaerobic to aerobic) which influenced the nitrification and denitrification rates. In the carbon section, the assimilation by papyrus was estimated on the basis of maximum photosynthetic rate and irradiance, and limited by the N concentration in the papyrus biomass. Other processes in the carbon section were mortality, fragmentation, leaching and harvesting. Biomass C and N content were linked through an optimum C:N ratio, so that changes in carbon (for example when harvesting occurs) led to proportional changes in the amount of N in the N section of the model. In the N section, mortality of papyrus biomass led to dead organic N which was further degraded to NH_4, which could be converted to NO_3 and enter denitrification depending on moisture conditions.

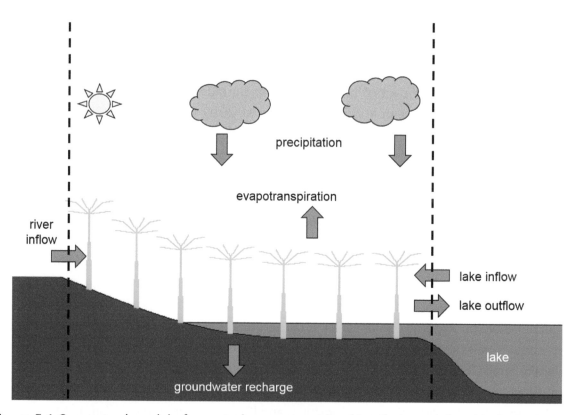

Figure 5.1 Conceptual model of a rooted papyrus wetland bordering a lake, the dashed lines indicate the model boundaries

In the current study, Papyrus Simulator was developed further in two areas: a phosphorus (P) section was added (1); and the model was generalized by merging the original seasonal and permanent wetland zones (based on Lake Naivasha) into one wetland zone that can be inundated based on the prevailing conditions of surface water inflow (a combination of stream and overland flows, in the model called river inflow), lake inflow (in the model this can be switched on and off depending on whether a seasonally inundated or permanently inundated wetland is simulated), and precipitation (2). All state and rate variables (processes) of the new model are given in Appendix 5.1 and 5.2, respectively, with their numbers between brackets in the text. All other parameters and variables are given in Appendix 5.3. The key processes and conceptual diagrams of the four sections (Figure 5.2a) are presented and described below.

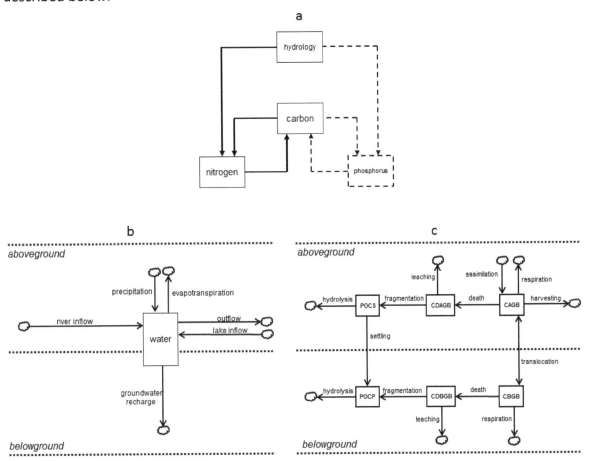

Figure 5.2 Model sections with - - - - new and → flow of information (a), conceptual diagram hydrology section with → flow of water (b) and conceptual diagram carbon section with → flow of carbon; POCS=particulate organic carbon in surface water; POCP= particulate organic carbon in pore water; CDAGB=carbon in dead aboveground biomass; CDBGB=carbon in dead belowground biomass; CAGB=carbon in aboveground biomass; CBGB carbon in belowground biomass (c)

5.2.2 HYDROLOGY

The hydrology section (Figure 5.2b) calculates the water level (as the volume of water per m^2 of surface area) based on inflow from river discharge, outflow to and backflow from the lake, precipitation, evapotranspiration and groundwater recharge. Backflow from the lake occurs when the sum of water outputs is greater than the sum of water inputs, keeping the surface water level constant at 0.5 m (72) and can be switched off to simulate wetlands that have an outflow, but do not receive backflow, allowing the surface water to drop in the dry season.

When the inputs are higher than the outputs the difference flows to the lake, keeping the surface water level at a maxmimum of 0.5 m (100). Lake Naivasha conditions (Gaudet 1979) were used as a basis to describe an annual discharge, precipitation and evapotranspiration regime. Groundwater recharge is modelled with first-order equations resulting in a maximum recharge of 6 mm day^{-1}, when soil is saturated (133). The factor MODE (based on van der Peijl and Verhoeven, 1999) is calculated as the proportion of soil porosity that is filled with water (Appendix 5.3). Ranging from 0 (soil completely saturated, anaerobic) to 1 (soil at field capacity or less, aerobic conditions), MODE influences nitrification and denitrification rates in the soil pore water. Conditions in the surface water are assumed to be aerobic. The soil depth considered in the model was 0.2 m.

5.2.3 CARBON

Papyrus biomass is modelled as carbon (C) in aboveground biomass (AGB), culms and umbels, and belowground biomass (BGB), rhizome and roots (Figure 5.2). Carbon in AGB (CAGB) results from net growth (assimilation minus mortality minus respiration) and translocation of C to and from the rhizome. Assimilation (42) was described using a logistic model depending on total aboveground biomass and limited by irradiance and the N and P concentrations in the plant, both described by Monod-type equations:

$$assimilation = max\_assimilation\_constant \times CAGB$$
$$\times \left(1 - \frac{CAGB}{C\_conc\_AGB * K\_assim}\right)\left(\frac{radiance}{radiance + K\_radiance}\right)$$
$$\times \left(\frac{limit\_N\_ass \times limit\_P\_ass}{0.81}\right)$$

in which max_assimilation_constant is the maximum relative assimilation rate (day^{-1}), CAGB is carbon in AGB (g C m^{-2}), C_conc_AGB is carbon in AGB (g C g^{-1} DW), K_assim is papyrus biomass at which assimilation stops (g DW m^{-2}), radiance is irradiance (MJ m^{-2} day^{-1}), K_radiance is the half saturation constant of irradiance for assimilation (MJ m^{-2} day^{-1}), limit_N_ass (-) and limit_P_ass (-) are limiting factors of N and P, respectively, for C assimilation (Appendix 5.3), the factor 0.81 ensures that maximum growth limitation can be reached (van der Peijl and Verhoeven 1999). At N or P concentrations in papyrus below the minimum needed for assimilation ($1.6 \cdot 10^{-3}$ g N g^{-1} DW and $8 \cdot 10^{-5}$ g P g^{-1} DW), no growth is assumed so a Monod-type function with a cut-off is used (van der Peijl and Verhoeven 1999). The dead AGB is fragmented and hydrolysed according to first order kinetics (48 and 113). Respiration in AGB (45) is the sum of maintenance respiration (proportional to biomass) and growth respiration (proportional to assimilation), as:

$$CAGB\_respiration$$
$$= CAGB \times maint\_coeff\_AGB + assimilation \times growth\_coeff\_AGB$$

in which maint_coeff_AGB is the maintenance respiration coefficient for AGB (day^{-1}), and growth_coeff_AGB is the respiration coefficient for assimilation (-).

Translocation of C between AGB and BGB (41) is based on an assumed optimal ratio between aboveground and belowground C (C_AGB_to_BGB_optimal_ratio, Appendix 5.3), of 1.2 (Boar et al. 1999; Jones and Humphries 2002). Whenever C ratio differs from this optimal ratio, C is translocated to restore the optimum:

$$if: C\_AGB\_to\_BGB\_optimal\_ratio < \frac{CAGB}{CBGB}$$

$$then: translocation = tldownmax\_C \times \left(\frac{CAGB/CBGB}{CAGB/CBGB + Ktldown\_C} \right)$$

$$if: C\_AGB\_to\_BGB\_optimal\_ratio > \frac{CAGB}{CBGB}$$

$$then: translocation = -tlupmax\_C \times \left(\frac{CBGB/CAGB}{CBGB/CAGB + Ktlup\_C} \right)$$

in which CAGB and CBGB are C in AGB and BGB, respectively (g C m^{-2}), C_AGB_to_BGB_optimal_ratio is the optimal ratio between C in AGB and BGB (-), tldownmax_C is the maximum rate for translocation of C from AGB to BGB (g C day^{-1}), tlupmax_C is the maximum rate for translocation of carbon from BGB to AGB (g C day^{-1}), Ktldown_C is the half saturation constant for downward translocation (-) and Ktlup_C is the half saturation constant for upward translocation.

Carbon in BGB (CBGB) is reduced by mortality following a first order equation depending on CBGB and a CBGB death constant (46). The dead BGB is fragmented and hydrolysed according to first order equations (50 and 112). Respiration by BGB (47) is calculated as for AGB, with growth respiration proportional to downward translocation and maintenance respiration proportional to CBGB.

5.2.4 NITROGEN

The N section (Figure 5.3) expresses the same aboveground and belowground alive and dead biomass compartments as the carbon model, expressed in g m^{-2} of N. Added to these are the main components of the N cycle in the wetland, particulate- and dissolved organic N, NH$_4$ and NO$_3$. Nitrate and ammonium are taken up by the BGB (84 and 94), and then passed on to the AGB by translocation (74). N in dead biomass is re-translocated (73) or passes through fragmentation (78 and 80), hydrolysis (116 and 119), mineralisation (56 and 59), nitrification (90 and 91) and denitrification (52).

Exchange of N between the aboveground and belowground layers takes place through diffusion (53, 82 and 92), driven by concentration differences of soluble compounds (NO$_3$, NH$_4$ and dissolved organic N), and settling of particulate organic N (115). The uptake of NH$_4$ and NO$_3$ by papyrus (84 and 94) depends on the carrying capacity for papyrus and the concentration of N in the biomass, and is limited by the concentration of NH$_4$ and NO$_3$, respectively. This limitation is modelled with a Monod-type equation, only the equation for NH$_4$ uptake is given here for illustration:

$$NH_4uptake = max\_NH_4\_uptake \times (N\_biomass) \times \left(1 - \frac{N\_biomass}{N\_max\_biomass}\right)$$
$$\times \left(\frac{NH_4P\_conc}{NH_4P\_conc + K\_NH_4}\right)$$

in which max_NH4_uptake is the maximum uptake rate of NH4 by papyrus (day[-1]), N_biomass is the amount of N in AGB and BGB (g N m[-2]), N_max_biomass the maximum amount of N stored in AGB and BGB (g N m[-2]), NH4P_conc is the NH4 concentration in the pore water (g N m[-3]), and K_NH4 is a half saturation constant (g N m[-3]). N_max_biomass (N_max_AGB + N_max_BGB) was calculated based on literature values for N content in both AGB and BGB, as 0.013 and 0.008 g N g[-1] DW, respectively (Appendix 5.3) and the maximum papyrus density of AGB and BGB found in literature, 8118 g DW m[-2] (Muthuri et al. 1989; Jones and Muthuri 1997).

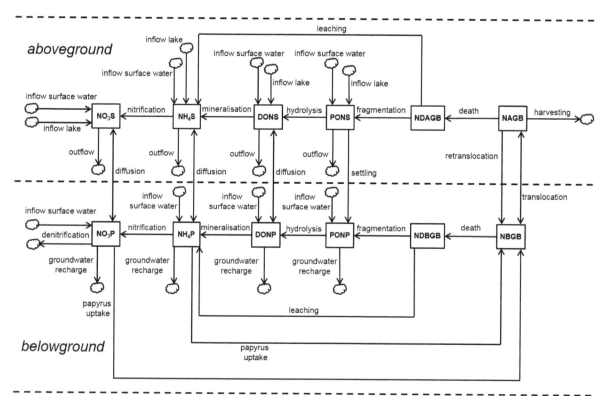

Figure 5.3 Conceptual diagram nitrogen section with → flow of nitrogen; NO3S=NO3 in surface water; NO3P=NO3 in pore water; NH4S=NH4 in surface water; NH4P=NH4 in pore water; DONS=dissolved organic N in surface water; DONP=dissolved organic N in pore water; PONS=particulate organic N in surface water; PONP= particulate organic N in pore water; NDAGB= N in dead AGB; NDBGB= N in dead BGB; NAGB= N in AGB; NBGB N in BGB

5.2.5 PHOSPHORUS

The P section is similar in structure to the N section (Figure 5.4), with the main exception being the adsorption (and release) of P to the sediment (31 and 32). These are modelled with a Langmuir equation and a factor that decreases adsorption and increases the release rate when the amount of adsorbed P approaches the maximum of adsorbable P (van der Peijl and Verhoeven 1999):

$$if: (OPADS < OPADS\_eq)$$
$$then: adsorption = \left(1 - \frac{OPADS}{OPADS\_max}\right) \times (OPADS\_eq - OPADS)$$

$$if: (OPADS > OPADS\_eq)$$
$$then: release = \frac{OPADS}{OPADS\_max} \times (OPADS - OPADS\_eq)$$

in which OPADS is the amount of P adsorbed (g P m^{-2}), OPADS_max is the maximum amount that can be adsorbed (g P m^{-2}) and OPADS_eq is the amount adsorbed in equilibrium with the orthophosphate (OP) concentration (g P m^{-2}).

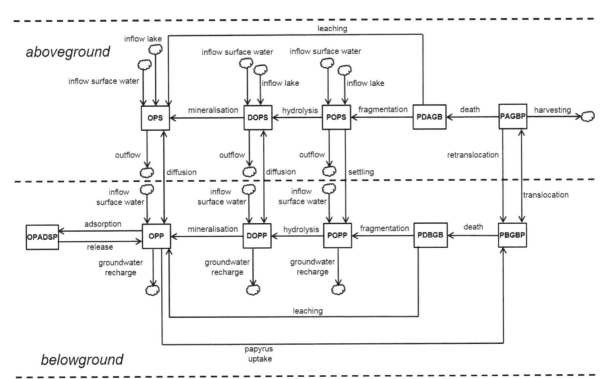

Figure 5.4 Conceptual diagram phosphorus section with → flow of phosphorus; OPADSP=orthophosphate adsorbed; OPS= orthophosphate in surface water; OPP= orthophosphate in pore water; DOPS=dissolved organic P in surface water; DOPP=dissolved organic P in pore water; POPS=particulate organic P in surface water; POPP= particulate organic P in pore water; PDAGB= P in dead AGB; PDBGB= P in dead BGB; PAGB= P in AGB; PBGB P in BGB

5.2.6 MODEL ASSUMPTIONS AND IMPLEMENTATION

It was assumed that the distribution of water in the wetland is uniform without preferential flow. Floating papyrus mats at the edge of the lake were not considered, because harvesting is assumed to take place in rooted papyrus zones. All elements needed for growth that are not included in the model were assumed not to be limiting. Uptake of NO_3, NH_4 and OP are not taking place when pore water is not available. The model currently does not include ammonia volatilization, N fixation and N deposition, considerations were discussed in Chapter 4.

The model was implemented in STELLA 10.0.6 (isee systems inc., Lebanon, NH, US) and run for a period of 5 years with rectangular (Euler) integration and a time step of 0.0625 days (1.5 h) for 1 m^2 of wetland. A complete listing of the equation layer of the Stella model is given in Appendix 5.4.

5.2.7 PARAMETERIZATION AND CALIBRATION

The model was parameterized and calibrated with data from the literature for Lake Naivasha (Appendix 5.1 and 5.3). When values from Lake Naivasha were not available, data from other East African papyrus wetlands were used. For parameters that were never studied or measured in papyrus wetlands, literature values from other wetland types were used or estimated (for details see Appendix 5.1 and 5.3). Seasonal variability (solar radiation and hydrological inputs) in the Lake Naivasha wetland was described using monthly averages from the period 1970-1982 for irradiance (Muthuri et al. 1989), and 1974-1976 for evaporation and precipitation (Gaudet 1979) and river inflow. Evaporation was multiplied with a factor 1.25 to estimate evapotranspiration for papyrus (Saunders et al. 2007). Monthly river inflow, based on the flow regime of the Malewa River (Gaudet 1979) was calibrated to achieve realistic flow rates and N and P concentrations. With a main rainy season in the months March-May and a short rainy season in December, evapotranspiration (4.1 - 6.2 mm day^{-1}) and precipitation (0.1 – 6.2 mm day^{-1}) varied throughout the year (Figure 5.5c). Two flooding conditions were simulated: in the permanently flooded wetland zone, it was assumed that there was a backflow from the lake of between 0 and 0.12 m^3 m^{-2} day^{-1} whenever river flow and rainfall were low. In the seasonally flooded zone, it was assumed that no lake backflow occurred, resulting in lower water levels during periods with low river flow and rainfall. This hydrological regime was repeated annually for the 5-year simulation (Figure 5.5).

5.2.8 RETENTION AND HARVESTING

Absolute N and P retention (g N or P m^{-2} yr^{-1}) were calculated as ($N_{inflow} - N_{outflow}$) and ($P_{inflow} - P_{outflow}$), both over the 5th year, when the model had stabilised (Figure 5.6a). N_{inflow} was the sum of N in river discharge (55, 58, 85, 87, 95, 97, 117 and 120) and N flowing in from the lake (54, 83, 93 and 114), and $N_{outflow}$ was the sum of N outflow to the lake (60, 88, 98 and 121) and groundwater recharge (57, 86, 96 and 118). P_{inflow} (34, 37, 38, 63, 66, 67, 123, 127 and 129) and $P_{outflow}$ (36, 39, 65, 69, 126 and 130) were calculated in the same way. Relative retention (% wt N or P m^{-2} yr^{-1}) was also calculated over the 5th year as ($N_{inflow} - N_{outflow}$)/N_{inflow} and ($P_{inflow} - P_{outflow}$)/P_{inflow}.

Two types of harvesting scenarios were applied (44, 76 and 104): daily harvesting in g papyrus DW m^{-2} d^{-1}; and annual harvesting in a percentage of standing biomass. The annual harvest took place during the dry season (on day 230) when the surface water levels go down (Figure 5.5a) and seasonal agriculture is most likely.

5.2.9 SENSITIVITY ANALYSIS

Initially, a "One at a Time" (OAT) local sensitivity analysis was done, by running the model with calibrated initial values of state variables and model parameters, and values of -10% and +10% of each of these. The effect of these variations on the output variables (papyrus biomass, nutrient concentrations in surface water, nutrient retention) were observed and 28 out of 81 parameters to which the model was most sensitive were selected for a global sensitivity analysis based on the approach outlined in Saltelli et al. (2000).

For each of the 28 parameters, a range of possible values (minimum, maximum) was determined, based on what was assessed as being most realistic (Table 5.1) and a rectangular distribution was set within the sensitivity settings of Stella. The model was then run 500 times, both for permanently and seasonally flooded conditions, with parameter values drawn from the 28 distributions for each run. As output variables, the following ten were selected: papyrus AGB, papyrus BGB (both in g DW m^{-2}), NH$_4$-N, NO$_3$-N and OP-P in the surface and pore water (in g N or P), and N and P retention for year 5 (in g m^{-2} y^{-1}). For the first eight variables, the mean value for the inundated period (days 32 to 322) of year 5 of the simulation run was computed. For N and P retention, the end values of year 5 were used. The resulting output dataset of two times 500 model runs with combinations of random parameters and output variables was submitted to multiple regression analysis, with the output variable as dependent variable. For each regression model that had a sufficiently high coefficient of determination (preferably $R^2 > 0.7$) standardized regression coefficients (or beta weights) were calculated and compared to assess the contribution of each input variable in explaining the variation in the output variable (as a measure of sensitivity of the output variables to the inputs). Only model parameters that had a significant regression coefficient (t-test, $P<0.05$) were included. All regression models were calculated using functions lm() and lm.beta() in R version 3.5.0 (R Core Team, 2018).

5.2.10 COMPARISON WITH FIELD DATA

To compare the simulation results with field data, papyrus wetland studies that reported the main output variables of the model (AGB and BGB, N and P concentrations in biomass, water quality, net primary production of biomass) were reviewed. Wetland characteristics, location, altitude and type (floating or rooted) were also included. As the model was aimed at papyrus wetlands in general and not at one wetland site in particular, model output was compared with the ranges of values found in the literature.

Table 5.1 Ranges of parameter values for global sensitivity analysis

variable	model value	min	max	unit	rationale
NH$_4$_concentration_river	3	0.03	6	g N/m^3	a
NO$_3$_concentration_river	2	0.02	4	g N/m^3	a
OP_concentration_river	0.5	0.005	1	g P/m^3	a
K_nitrification	0.005	0.00005	0.01	day^{-1}	a
K_denitrification	0.01	0.0001	0.02	day^{-1}	a
OPADS_maxdw	0.004	0.00004	0.008	g P/g DW	a
Porosity	0.8	0.6	1	-	b
K_NH$_4$_diffusion	0.1	0.001	0.2	m^2/day	a
K_NO$_3$_diffusion	0.1	0.001	0.2	m^2/day	a
K_OP_diffusion	0.1	0.001	0.2	m^2/day	a
max_NH$_4$_uptake	0.1	0.001	0.2	day^{-1}	a
max_NO$_3$_uptake	0.1	0.001	0.2	day^{-1}	a
max_OP_uptake	0.1	0.001	0.2	day^{-1}	a
papyrus_max_biomass	8118	6494.4	9741.6	g DW/m^2	b
max_assimilation_constant	0.17	0.085	0.255	day^{-1}	c
K_assimilation	5150	4120	6180	g DW/m^2	b
CAGB_death_constant	0.0057	0.000057	0.0114	day^{-1}	a
CBGB_death_constant	0.0014	0.000014	0.0028	day^{-1}	a
CDAGB_fragmentation_constant	0.0005	0.000005	0.001	day^{-1}	a
CDBGB_fragmentation_constant	0.00083	0.000008	0.00166	day^{-1}	a
CDABG_leaching_constant	0.432	0.00432	0.864	-	a
CDBGB_leaching_constant	0.486	0.00486	0.972	-	a
N_retranslocation_constant	0.7	0.007	1.4	-	a
P_retranslocation_constant	0.77	0.0077	1.54	-	a
NAGBlitaverage	44	35.2	52.8	g N/m^2	b
NBGBlitaverage	31	24.8	37.2	g N/m^2	b
PAGBlitaverage	2.61	2.088	3.132	g P/m^2	b
PBGBlitaverage	2.78	2.224	3.336	g P/m^2	b

a = minimum value is 1% of original value and maximum value is 200% of original value; for parameters that likely vary from almost 0 to double the value used in the model;

b = minimum value is 80% of original value and maximum value is 120% of original value; for parameters with model values that were not likely to differ from reality;

c = minimum value is 50% of original value and maximum value is 150% of original value; for parameters that would never be close to zero, but may be different from reality

5.3 RESULTS

5.3.1 WATER LEVELS, IN- AND OUTFLOWS AND FACTOR MODE

Under permanently flooded conditions, the surface water level was constant at 0.5 m above the substrate (Figure 5.5a), resulting in a value of 0 for the controlling factor MODE (Figure 5.5b), implicating saturation (and anaerobic conditions) in the pore water. Under seasonally flooded conditions, the surface water level was 0.5 m during the rainy season and dropped to zero in the dry season while the pore water was below saturation for part of the dry season (Figure 5.5a). This resulted in peaks of the MODE factor (Figure 5.5b). In the dry season, when the sum of river inflow and rainfall was smaller than the combined evapotranspiration and recharge , the seasonal wetland did not receive backflow from the lake while the permanent wetland did (Figures 5.5d and e).

5.3.2 ABOVE- AND BELOWGROUND BIOMASS AND EFFECTS OF HARVESTING

The simulated papyrus biomass reached a maximum of 8127 g DW m⁻² in year 5, when the model was assumed stable. Belowground biomass (BGB) remained higher than aboveground biomass (AGB, Figures 5.6a and b). Without harvesting and under permanently flooded conditions, AGB and BGB fluctuated annually between 3809 and 3824 g DW m⁻² and 4287 and 4304 g DW m⁻², respectively. With seasonally flooded conditions, fluctuation was slightly higher: 3798 to 3823 g DW m⁻² (AGB) and 4276 to 4303 g DW m⁻² (BGB). Once per year there was a small drop in biomass coinciding with the dry season (Figures 5.5a and 5.6b). The AGB:BGB ratio remained similar at 0.9 throughout the year in both zones.

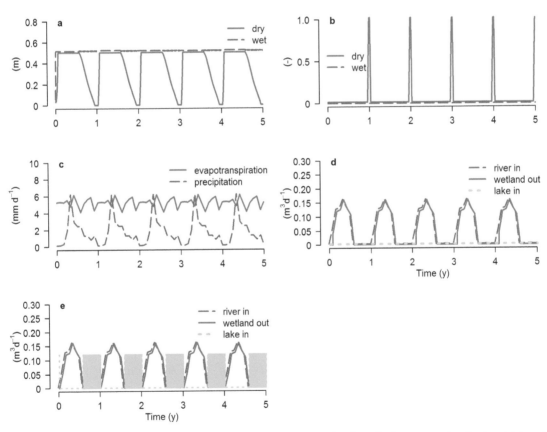

Figure 5.5 Simulated hydrology: water level permanently flooded system in dashed blue line and seasonally flooded system in solid red line (a) factor describing saturation (MODE) permanently flooded system in dashed blue line and seasonally flooded system in solid red line (b) evapotranspiration in solid red line and precipitation in dashed blue line (c) river inflow in dashed blue line, wetland outflow in solid red line and lake inflow in dotted green line for seasonally flooded system (d) and river inflow in dashed blue line, wetland outflow in solid red line and lake inflow in dotted green line for permanently flooded system (e)

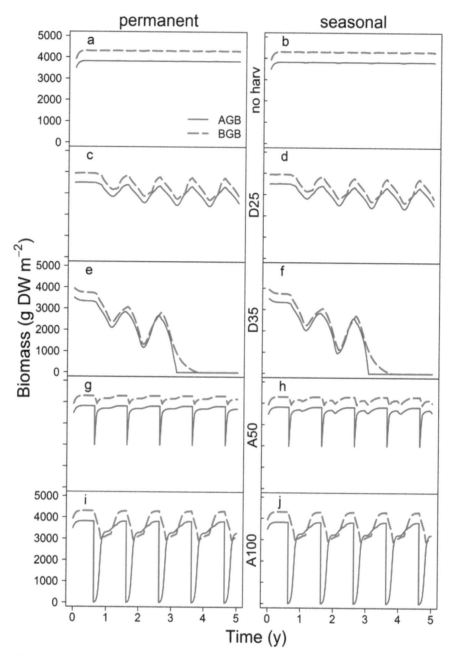

Figure 5.6 Simulated biomass with different aboveground biomass harvesting scenarios: no harv=no harvesting; D25=harvest of 25 g DW m⁻² d⁻¹; D35=harvest of 35 g DW m⁻² d⁻¹; A50=harvest of 50% AGB once per year; A100=harvest of 100% AGB once per year

When a harvesting rate of 25 g DW m⁻² d⁻¹ of AGB was applied, the negative impact of the dry season on AGB and BGB increased and total biomass was reduced by 13-37% in the permanent zone and 14-39% in the seasonal zone (Figures 5.6c and d). If harvesting was increased to 35 g DW m⁻² d⁻¹, the AGB was reduced to zero after only 3 years in the seasonally flooded wetland and about 20 days later in the permanently flooded wetland. With a one-time annual harvest of 50% or 100% the biomass recovered fully in about 11 months. In contrast with the daily harvesting scenarios, the AGB:BGB ratio changed during the recovery period, increasing from very low just after the harvest to the original ratio (0.9) after recovery.

For the 100% harvest the AGB:BGB ratio increased to more than unity before returning to the original value (Figures 5.6i and j). This was a result of temporary slow recovery of BGB owing to nutrient limitation during the absence of river inflow (Figures 5.5d, 5.5e, 5.7i and 5.7j).

5.3.3 EFFECTS OF HARVESTING ON SURFACE WATER CONCENTRATIONS OF NO_3, NH_4, AND OP

Under permanently flooded conditions, the concentrations in year 5 of NO_3-N, NH_4-N and OP-P in the surface water reached a peak of 1.6, 3.0 and 0.5 g m^{-3}, respectively (Figure 5.7a). These peaks coincided with the wet season (Figure 5.5). The lowest concentrations in year 5 of zero (NO_3-N), 0.2 (NH_4-N) and zero (OP) g m^{-3} occurred at the end of the year, just before the start of the wet season and at the end of a period with low N and P inputs from river inflow (Figures 5.5d, 5.5e and 5.7a). The patterns in the permanently flooded and seasonally flooded systems were similar, as were maximum concentrations recorded. The lowest concentrations were close to zero due to the absence of input of N and P from the lake during the dry seasons (Figure 5.7b).

With a harvesting rate of 25 g DW m^{-2} d^{-1} under permanently flooded conditions, the highest concentrations of NO_3-N (0.8 g m^{-3}), NH_4-N (1.2 g m^{-3}) and OP (0.2 g m^{-3}) dropped and occurred earlier after the start of the dry season compared with no harvesting (Figure 5.7c). For seasonally flooded conditions, nutrient concentrations were similar (Figure 5.7d). When harvesting was increased to 35 g DW m^{-2} d^{-1} under permanently flooded conditions, the minimum concentration in year 5 for all compounds were higher than without harvesting (NO_3-N 2.0; NH_4-N 2.3; and OP 0.4, all g m^{-3}). Maxima of NO_3 (3.2 g N m^{-3}) and OP (0.5 g P m^{-3}) occurred at the end of year 5. Ammonium maxima (3.1 g N m^{-3}) occurred early in the year. Under seasonally flooded conditions, the effects on the lowest concentrations were the same as under permanently flooded conditions for NO_3 and OP, but higher for NH_4 (3.0 g N m^{-3}). The highest concentrations for NO_3 and NH_4 were around 5 g N m^{-3} just before and after the dry periods when surface water dropped (Figures 5.7e and 5.7f).

When 50% of the AGB was harvested under both hydrological conditions, all nutrient concentrations dropped from the moment of harvesting to the point of biomass recovery (Figures 5.6g, 5.6h, 5.7g and 5.7h). With 100% harvesting, the trend was similar under both seasonally and permanently flooded conditions. Just after the harvest there was an increase in all concentrations, which was slightly lower under seasonally flooded conditions due to the absence of backflow (with nutrients) from the lake. After this increase, concentrations decreased and remained lower compared with the situation without harvesting, until the biomass recovered (Figures 5.6i, 5.6j, 5.7i and 5.7j).

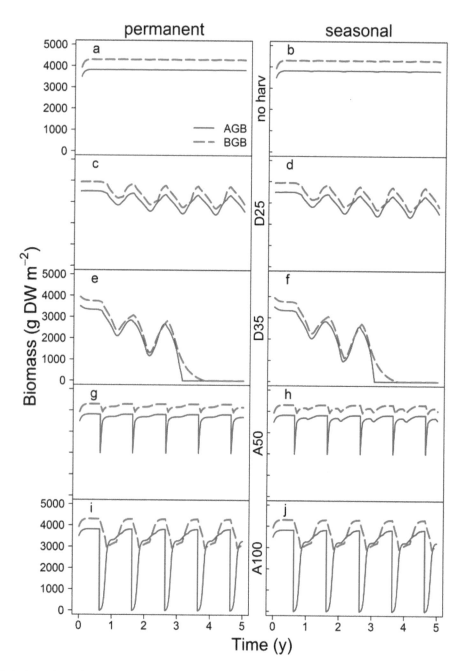

Figure 5.7 Simulated inorganic N and P concentrations with different aboveground biomass harvesting scenarios: no harv=no harvesting; D25=harvest of 25 g DW m⁻² d⁻¹; D35=harvest of 25 g DW m⁻² d⁻¹; A50=harvest of 50% AGB once per year; A100=harvest of 100% AGB once per year

5.3.4 EFFECTS OF HARVESTING ON N AND P RETENTION

N and P retention increased with increasing harvesting rates in both permanently and seasonally flooded conditions until a dramatic drop when the papyrus was over-harvested (Figures 5.6e, 5.6f and 5.8). For the permanently flooded wetland, this point for both N and P was at a harvesting rate of about 34 g DW m⁻² d⁻¹. In seasonally flooded conditions, this was similar at around 33 g DW m⁻² d⁻¹. N retention under both flooding conditions, ranged from 10 g N m⁻² yr⁻¹, without harvesting, to 67 g N m⁻² yr⁻¹ at a harvesting rate of 33 g DW m⁻² d⁻¹, dropping to a net maximum release of 3.9 g N m^{-2} yr^{-1} under permanently flooded conditions. Under seasonally flooded conditions, N retention was 12 g N m^{-2} yr^{-1} without harvesting, increasing to 66 g N m^{-2} yr^{-1} and then falling to a release of 3.4 g N m^{-2} yr^{-1} at higher harvesting rates. P retention increased from 0.6 to 6.3 and from 0.6 to 6.2 g P m^{-2} yr^{-1} in permanently and seasonally flooded conditions, respectively, thereafter declining with faster harvesting to around zero for both conditions.

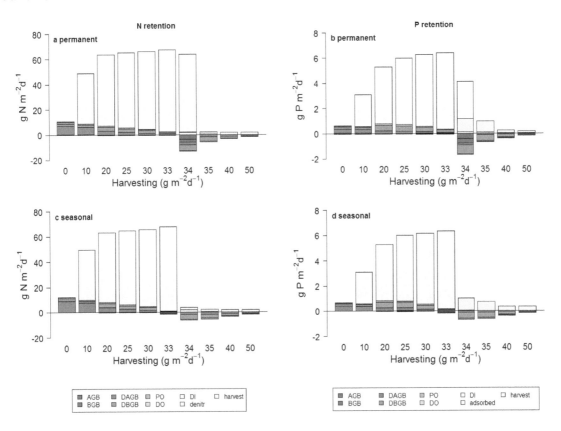

Figure 5.8 Nitrogen and phosphorus retention in a seasonal and permanently flooded system with different harvesting rates

Without harvesting (Figure 5.8), N and P were retained mainly in accumulating dead AGB (7.1 g N m^{-2} yr^{-1} and 0.3 g P m^{-2} yr^{-1}) and (8.7 g N m^{-2} yr^{-1} and 0.4 g P m^{-2} yr^{-1}) for permanent and seasonal systems, respectively. To a lesser extent, N and P were also retained in dead BGB and in particulate and dissolved organic matter. With increased harvesting rates the accumulation in organic matter (N and P) gradually decreased, while overall retention increased as a consequence of uptake by recovering papyrus (Figure 5.8). Denitrification was marginal in both zones until papyrus was overharvested, when it increased to 1.8 g N m^{-2} yr^{-1} (permanent) and 2.0 g N m^{-2} yr^{-1} (seasonal) and became the sole process responsible for retaining, or more correctly, removing N (Figures 5.8a and c). The denitrification was slightly higher under seasonal conditions due to the higher NO$_3$ concentrations (Figures 5.7e and f) compared with permanent flooding conditions and the short aerobic period under seasonally flooded conditions (Figure 5.5b). At harvesting rates leading to the absence of papyrus (Figures 5.6e and f) the main factor for P retention was adsorption, at 1.0 g P m^{-2} yr^{-1} in both systems (Figures 5.8b and d). The values taken are from the 34 g DW m^{-2} d^{-1} harvesting scenario. When harvesting was increased further, adsorption rates were lower.

Table 5.2 Adjusted R^2 values of regression models of dependent variables

dependent variable	permanently flooded system	seasonally flooded system
AGB	0.77	0.75
BGB	0.71	0.70
NH$_4$S	0.82	0.79
NH$_4$P	0.75	0.72
NO$_3$S	0.84	0.81
NO$_3$P	0.72	0.70
OPS	0.94	0.92
OPP	0.85	0.84
N retention	0.58	0.60
P retention	0.53	0.54

5.3.5 SENSITIVITY ANALYSIS

Table 5.2 shows the results of the twenty regression models and the ten output (dependent) variables for both permanently flooded and seasonally flooded systems. All models were significant, but the four models for N and P retention had adjusted R^2 below 0.7 (0.53 - 0.60). Despite this, all models where used for further analysis.

5.3.5.1 SENSITIVITY ANALYSIS FOR BIOMASS OUTPUTS

There was little difference between responses of biomass under permanently and seasonally flooded conditions and between AGB and BGB (Figures 5.9a and b). Biomass responded most positively to an increase in the maximum assimilation constant and the K assimilation, with beta values between 0.46 and 0.55. For both constants, responses where higher for permanent than for seasonally flooded conditions, and for AGB than for BGB. Positive responses (beta between 0.10 and 0.15) were also observed with increased NH$_4$ and NO$_3$ concentrations in the inflow for AGB and BGB and for N translocation constant on BGB for both conditions. Biomass responded negatively to an increased aboveground death constant (beta between -0.35 and -0.41), with the responses to BGB more negative than to AGB, and also slightly more negative for seasonal than for permanently flooded conditions.

For N (Figures 5.9c, d, e and f) the differences between dry and wet conditions were modest. The concentrations in the incoming river water had the highest impact on the concentrations in both wetland systems, especially on surface water concentrations. For surface water, NH$_4$ river concentrations explained the NH$_4$ concentration in the surface water with a beta value of 0.85 and 0.79 for permanently flooded and seasonally flooded conditions, respectively. For the pore water concentrations the beta values were lower, 0.69 and 0.65. The incoming NO$_3$ concentrations positively influenced NO$_3$ surface water concentrations in both systems (beta=0.77 for permanent) and (beta=0.74 for seasonal), and in the pore water beta values of

0.62 and 0.60, respectively. After the river concentrations, the NH₄ and NO₃ concentrations were most sensitive to changes in the N re-translocation constant. Beta values for the re-translocation constant influencing NO₃ concentrations were higher than for NH₄ and re-translocation beta values in pore water for both NO₃, and NH₄ concentrations were higher than those in surface water (Figures 5.9c, d, e and f). There were also positive responses (betas around 0.20) of the NH₄ concentration in surface and pore water in both systems to an increased belowground leaching constant. Higher soil porosity (dilution) resulted in lower NO₃ and NH₄ concentrations in the pore water with betas around -0.15. Increased death constants for AGB and, more so, for BGB had a negative effect on both N species, but more on NO₃ than NH₄ (Figure 5.9c, d, e and f).

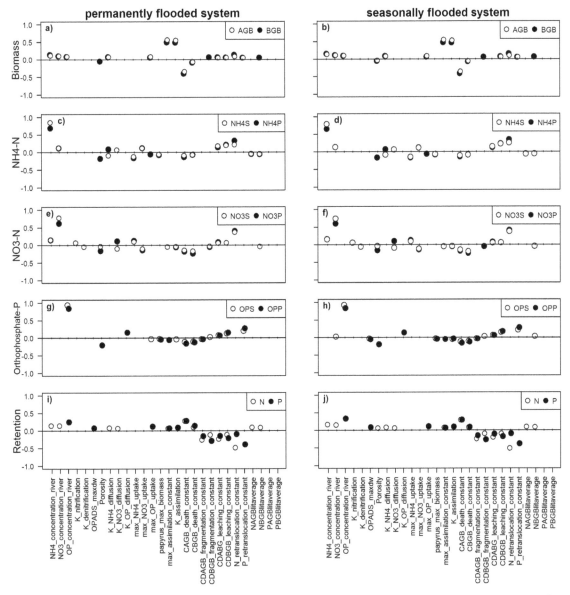

Figure 5.9 Beta values with dependent variables on the y-axis and input parameters on the x-axis, with: NH4S=NH4-N in surface water; NH4P=NH4-N in pore water; NO3S=NO3-N in surface water; NO3P=NO3-N in pore water; OPS=OP-P in surface water; OPP=OP-P in pore water

For P, there was also not much difference in sensitivity between the permanently and seasonally flooded systems (Figures 5.9g and h). The OP concentrations in the wetland were most sensitive to the river concentrations, for surface water with beta values of 0.94 (permanent) and 0.92 (seasonal) and pore water 0.84 (permanent) and 0.83 (seasonal). The P re-translocation constant explained OP in pore water with beta of 0.28 for both systems and in surface water with beta=0.20 (permanent) and beta=0.22 (seasonal). Similar to the N concentrations also the OP concentrations in the pore water were sensitive to changes in porosity (beta 0.20 for both systems). Like NH_4, the OP concentrations in both surface and pore water increased with increasing belowground leaching constant (beta values around 0.15). Similar to NH_4 and NO_3, the concentrations of OP decreased with higher death constants (beta values between 0.11 and 0.16).

5.3.5.2 SENSITIVITY ANALYSIS FOR NITROGEN AND PHOSPHORUS RETENTION

N retention was largely sensitive to the same parameters in both systems. The most influential factor was the N re-translocation constant (beta values -0.48 and -0.52) for the permanently and seasonally flooded systems, respectively (Figures 5.9i and j). Higher aboveground fragmentation and leaching constants led to lower N retention: beta values -0.26 and -0.23, respectively (permanent) and -0.25 and -0.20 (seasonal). N retention increased with a higher aboveground death constant 0.28 (permanent) and 0.29 (seasonal). Increasing inflow NH_4 and NO_3 concentrations increased retention in both systems (beta values 0.14 - 0.16).

P retention (Figures 5.9i and j) was mainly sensitive to the OP concentration in the influent (beta 0.25 permanent, 0.33 seasonal) and, similar to N, the P re-translocation constant (beta -0.39 permanent and -0.38 seasonal) and the aboveground death constant (beta 0.28 permanent and 0.29 seasonal). P retention was also sensitive to increase in fragmentation and leaching constants (beta -0.28 and -0.21 permanent and -0.26 and -0.18 seasonal). However, in contrast to N retention of mainly BGB and much less of AGB (Figure 5.9i and j), this was likely a consequence of the higher re-translocation constant for P (Appendix 5.3).

5.3.6 N:P RATIO IN THE SURFACE WATER

The N:P ratio coming into the wetland was 9.9 for permanent and 10.0 for seasonally flooded conditions (Table 5.3). Without harvesting and with sustainable harvesting rates (D25, A50 and A100), the TN:TP ratio was reduced to values between 8.2 (A100) and 9.6 (no harvesting) for both flooding conditions, indicating that relatively more N than P was retained. With overharvesting (D35), the TN:TP ratio increased to 10.3. The main change occurred with dissolved inorganic N and P, DIN:DIP ratios dropped between 6.4 and 8.9 (Table 5.3). The particulate N to P ratio increased with no harvesting and annual harvesting (A50 and A100) and did not change much with daily harvesting (D25) and over-harvesting (D35), while the ratio of dissolved organic compounds was hardly affected by the wetland (Table 5.3).

Table 5.3 Nitrogen to phosphorus ratios in water in inflow and outflow

	system	in	out 0	out D25	out D35	out A50	out A100
TN:TP	P	9.9	9.6	9.4	10.2	9.0	8.2
TDN:TDP	P	9.9	9.1	9.2	10.4	8.4	7.4
DIN:DIP	P	9.9	8.9	8.8	10.5	7.9	6.5
DON:DOP	P	9.9	10	9.8	9.8	10	9.9
PON:POP	P	10.0	11.5	9.7	9.7	11.2	10.5
TN:TP	S	10.0	9.6	9.4	10.3	9.0	8.2
TDN:TDP	S	10.0	9.1	9.2	10.4	8.3	7.4
DIN:DIP	S	10.0	8.9	8.8	10.5	7.8	6.4
DON:DOP	S	10.0	10.2	9.9	9.9	10.1	10.1
PON:POP	S	10.0	11.6	9.8	9.8	11.2	10.6

With TN=Total N; TP=Total P; TDN=Total Dissolved N; TDP=Total Dissolved P; DIN=Dissolved Inorganic N; DIP=Dissolved Inorganic P; DON=Dissolved Organic N; DOP=Dissolved Organic P; PON=Particulate Organic N; POP=Particulate Organic P; P=permanently flooded system; S=seasonally flooded system; in=inflow; out 0=outflow no harvesting; out D25=outflow 25 g DW m^{-2} d^{-1} harvesting; out D35=outflow 35 g DW m^{-2} d^{-1} harvesting; out A50=outflow 50% once a year harvesting and out A50=outflow 100% once a year harvesting. All ratios are weight ratios.

5.3.7 COMPARISON WITH FIELD DATA

Since the 1970s frequent measurements have been made across a range of wetlands on biomass, water quality and productivity (Table 5.4). A wide range of AGB and BGB values were reported in studies ranging from South Africa to Egypt and from sea level in the Nile Delta to 1883 m altitude at Lake Naivasha. Maximum values were higher in rooted systems (Table 5.4). Water quality values vary from low concentrations for e.g. Lake Naivasha and the Shire River in Malawi (Gaudet, 1975) in the 1970s to high concentrations in wetlands with wastewater input like Namiiro, Uganda (Kipkemboi et al. 2002) and Nakivibo, Uganda (Kansiime er al. 2007). Unfortunately, productivity values were reported from studies that did not include water quality data and could not be related to nutrient inputs. Productivity values above 20 g m^{-2} d^{-1} were reported at altitudes ranging from 700 (Upemba, DR Congo) to 1883 at Lake Naivasha (Table 5.4). Values simulated with the model for AGB and BGB, N and P concentrations in biomass, water quality and net primary production of biomass were all well within the ranges found in the literature (Table 5.4).

5.4 DISCUSSION AND CONCLUSION

5.4.1 MODEL PERFORMANCE

The current model focussed on N and P processes related to papyrus vegetation (growth, mortality, nutrient uptake and release), but also included microbiological and physico-chemical processes. The model studied the impact of different hydrological (permanent and seasonal flooding) and harvesting regimes on N and P retention. The hydrology section enabled studying the effect of hydrological regimes independently, for example by including or excluding backflow from the lake. The development of the P model enabled studying retention at process level in comparison with N, and simulating changes in N:P ratios in the water from inflow to outflow. Comparison of model outcomes with results of field studies (Table 5.4) confirmed that the model produced reasonable estimates of biomass, N and P in biomass, productivity and the concentrations of nutrients in the water. The simulated time of 11 months for re-growth of AGB (Figures 5.6i and 5.6j) was realistic compared with literature values of 6-12 months (Muthuri et al. 1989; Kansiime and Nalubega 1999; Terer et al. 2012b).

5.4.2 N AND P RETENTION

N and P retention were mainly a result of accumulation in dead biomass (Figure 5.8). Both relative and absolute P retention (4 wt% m^{-2} yr^{-1} and 0.6 g P m^{-2} yr^{-1}) were lower than N retention (7 wt% m^{-2} yr^{-1} and 11 g N m^{-2} yr^{-1}). The absolute retention was lower due to lower P content in dead biomass and not compensated by P adsorption due to the low simulated P concentrations in the pore water. Gaudet (1977) also identified peat accumulation as the main mechanism for retention of N and P in floating papyrus, but found higher values (65 g N m^{-2} and 0.7 g P m^{-2} yr^{-1}), indicating either a higher mortality or lower decomposition rate than used for model simulation. Relative N retention also exceeded P retention. This was caused by a relatively high amount of N (N:P ≈ 20) in dead AGB compared with living AGB (N:P ≈ 15) and, to a lesser extent, denitrification. The relatively low amount of P in AGB was caused by a higher re-translocation constant (0.77 for P vs 0.7 for N). Higher resorption of P under P limiting conditions is not exceptional for tropical wetland macrophytes, as shown for *Eleocharis cellulosa* and *Typha domingensis* in Belize (Rejmánková 2005; Rejmánková and Snyder 2008), but this remains to be confirmed for *Cyperus papyrus* dominated wetlands. Under the environmental conditions at Lake Naivasha, simulated papyrus growth was not P limited. Nevertheless, due to N fixation (not currently modelled) in the root zone (Gaudet 1979; Boar et al. 1999), P may still be the limiting nutrient in reality justifying a higher P re-translocation. Given the considerable impact of re-translocation on retention of N and P (Asaeda et al. 2008), as confirmed by the sensitivity analysis (Figure 5.9), empirical research into re-translocation rates in *C. papyrus* is recommended.

5.4.3 EFFECT OF MOISTURE ON NUTRIENT CYCLING

Differences in moisture conditions in the wetland (permanent soil saturation and standing water under permanent flooding, unsaturated soil and absence of water during part of the year under seasonal flooding, Figure 5.5) led to differences in papyrus biomass, nutrient retention and water quality. Under seasonal flooding, biomass was lower than under permanent flooding (Figure 5.6). The low water level led to both N and P limitation in the model, less uptake belowground and less translocation of N and P to the aboveground parts of the plant. The difference in impact on AGB compared with BGB was amplified by re-translocation of N and P resulting from AGB mortality. The dependence of re-translocation on water depth is known from *Eleocharis sphacelata* (Asaeda et al. 2008). With harvesting the impact of reduced water levels on papyrus biomass increased (Figure 5.6). Seasonally flooded conditions led to higher retention (Figure 5.8) and lower NO_3, NH_4 and OP concentrations in the water (Figure 5.7) than permanent flooding, due to the higher mortality in the dry season and the fast recovery during the wet periods of the year. This leads to a higher net N and P uptake (lower concentrations) and a larger accumulation of dead AGB (higher retention). The effects are small with the two hydrological scenarios presented here, but likely of greater importance with longer dry periods and lower concentrations of N and P in the inflow. While the temporary nutrient limitation in the dry season leads to a higher simulated net annual retention, in reality this may be less. Under water stress, re-translocation would likely increase, resulting in lower concentrations of N and P in AGB (Asaeda et al. 2008), and decomposition and N and P release would increase because of more aerobic conditions. Denitrification would also be lower with a longer dry period. Under the current seasonal flooding regime there was little difference (Figure 5.8a and 5.8c) and with over-harvesting denitrification was even higher under seasonal conditions as a result of a higher NO_3 concentration (Figures 5.6e, 5.6f, 5.8a and 5.8c).

5.4.4 EFFECTS OF HARVESTING

Harvesting increased both N (6 times) and P retention (10 times) and decreased the TN:TP mass ratio of the water from 10 to 8. This is in line with findings on N and P retention in a *C. papyrus* dominated wetland receiving wastewater runoff (Kanyiginya et al. 2010). As total N was retained more than total P (Figure 5.8), the TN:TP ratio in the water decreased. This effect was stronger with harvesting, however with overharvesting retention decreased and N:P ratio increased (Figure 5.8 and Table 5.3). The ratio of dissolved organic N to dissolved organic P did not change much, and the ratio for particulate N to particulate P in the water was even higher in the outflow. Reduction in TN:TP in the water, therefore, could be attributed to dissolved inorganic N:P ratio. Particulate N:P in the water increased due to mortality because N content in biomass is higher than the P content and therefore more particulate N is released in the outflow. Harvesting reduced the absolute mortality rate, because of a reduction in biomass (Figure 5.6) and, consequently, the particulate N:P in the wetland outflow. The reduction of the inorganic N:P ratio in the water is caused by the uptake of papyrus plants to replace vegetation that has died-off. With harvesting, this increased as the demand for inorganic N and P increases for re-growth and fast translocation from rhizome to AGB. As the uptake of N is higher than P, the N:P ratio in the water is reduced. A higher N than P retention was also found in field experiments (Gaudet 1977, Kansiime et al. 2007, Kanyiginya 2010). With over-harvesting, adsorption of OP increased as a result of higher OP concentrations in the pore water (Figures 5.7e, f and 5.8b and d), explaining the higher N:P ratio in the outflow compared with inflow (Table 5.3). Intact wetlands would therefore reduce both N and P and push the system to be more N limited.

Table 5.4 *Papyrus wetland field data compared with simulation results*

Wetland	country	altitude [m]	type[a]	agb [g m⁻²]	bgb [g m⁻²]	N total [%dw]	P total [%dw]	NO₃ [g N m⁻³]	NH₄ [g N m⁻³]	OP [g P m⁻³]	Productivity [g m⁻² day⁻¹]
Nkome[1]	Botswana	1130	r	3933	4971						
Upemba[2]	DR Congo	700	r	5173		0.65					13.7-25.9
Damietta Sharabas[3]	Egypt	0	r	3500-8000	2000-7500			0.44	0.46	0.025	
Kafr El-Batikh[3]	Egypt	0	r	2500-8000	2500-8500			0.44	0.45	0.026	
Naivasha[4]	Kenya	1883	r	3272	4516						17.2
Naivasha[5]	Kenya	1883	r	4652							
Naivasha[6]	Kenya	1883	r	3245							
Naivasha[7]	Kenya	1883	r			0.7	0.13				
Winam Gulf[9]	Kenya	1135	r	8457							
Busoro[6]	Rwanda	1350	r	1384							
iSimangaliso[10]	South Africa	30	r	3620							
Rubondo[11]	Tanzania	1150	r	8691	6398						
Akika island[2]	Uganda	915	r	5000							29-34.1
Gaba[12]	Uganda	1140	r	3875							5.1
Jinja[13]	Uganda	1140	r	2260	7130						22.1
Kirinya[14]	Uganda	1140	r	3290							
Kirinya[15]	Uganda	1140	r								
Lubigi[16]	Uganda	1170	r								16.7-37.4
Lubigi[22]	Uganda	1170	r	2700	900			0.02	2	0.2	
Mpigi[4]	Uganda	1150	r	2062							
Mpigi[22]	Uganda	1150	r	2000	700			0.05	2	0.15	
Nakivubo[14]	Uganda	1140	r	2480							
Nakivubo[15]	Uganda	1140	r						5.6		
Nakivubo[2]	Uganda	1140	r	2900							24.7
Congo River[17]	DR Congo	300	f			1.5	0.08	0.5	1.45	0.05	
Tchike[17]	DR Congo	10	f			0.9	0.074	0.01	1.3	0.135	

Wetland	country	altitude [m]	type[a]	agb [g m⁻²]	bgb [g m⁻²]	N total [%dw]	P total [%dw]	NO₃ [g N m⁻³]	NH₄ [g N m⁻³]	OP [g P m⁻³]	Productivity [g m⁻² day⁻¹]
Lake Tana[17]	Ethiopia	1788	f			0.79	0.062	0.05	3	0.21	
Naivasha[18]	Kenya	1883	f	6045	5495						
Naivasha[17]	Kenya	1883	f			1.33	0.052	0.02	0	0.065	
Naivasha[19]	Kenya	1883	f						0.28		
Naivasha[20]	Kenya	1883	f	3602							14.1-21
Shire River[17]	Malawi	475	f			0.78	0.049	0.02	0.1	0.006	
Pamplemousses[17]	Mauritius	100	f			1.05	0.024	0.22	0.07	0.12	
Gogonya[22]	Uganda	1135	f	2800	1400			0.1	2	0.2	
Kabanyolo[17]	Uganda	1190	f			1.75	0.049	0.3	1.6	0.06	
Lake George[17]	Uganda	915	f			1.28	0.099	0.04	4	0.34	
Lake George[21]	Uganda	915	f								
Mayanja[17]	Uganda	1075	f			0.87	0.039	0.3	0.64	0.01	
Nakivubo[22]	Uganda	1140	f	4000	1100			0.2	7	0.4	
Namiiro[22]	Uganda	1150	f	6000	1400			0.2	8	0.8	
Victoria[17]	Uganda	1135	f			1.11	0.053	0.001	0.4	0.01	
Various[21]		915	f			1.22	0.059				
Naivasha[8]	Kenya	1883	c	3605	3340						
Range (rooted)		0-1883	r	1384-8691	700-8500	0.65-0.7	0.13	0.02-0.44	0.45-5.6	0.025-0.2	5.1-37.4
Range (floating)		10-1883	f	2800-6045	1100-5495	0.78-1.75	0.024-0.099	0.001-0.5	0-8	0.006-0.8	14.1-21
Papyrus simulator		1883	r	3798-3824	4267-4304	0.9-1.0	0.070-0.072	0.0-1.6	0.2-3	0.0-0.5	30.3

r = rooted; f = floating; c = combined 1) Kiwango 2013; 2) Thompson et al. 1979; 3) Serag 2003; 4) Jones and Muthuri 1997; 5) Terer et al. 2012b; 6) Jones and Muthuri 1985; 7) Gaudet and Muthuri 1981; 8) Boar 2006; 9) Osumba et al. 2010; 10) Adam et al. 2014; 11) Mnaya 2007; 12) Kagwa et al. 2001; 13) Saunders et al. 2007; 14) Mugisha et al. 2007; 15) Kansiime et al. 2007; 16) Opio et al. 2014; 17) Gaudet 1975; 18) Boar et al. 1999; 19) Muthuri and Jones 1997; 20) Muthuri et al. 1989; 21) Gaudet 1977; 22) Kipkemboi et al. 2002

The TN:TP outflow ratio without harvesting was 9.6 g/g, equivalent to a 21.3 N:P molar ratio, and with harvesting this decreased to 18 N:P molar ratio. This is in between values found for Lake Victoria (13.6 N:P molar) and Lake Malawi (28.4 N:P molar) (Guildford and Hecky 2000). A TN:TP molar ratio in the water below 20 is considered N limiting and favours blooms of N fixing cyanobacteria at high P concentrations as they outcompete non-N fixing algal species (Guildford and Hecky 2000). Wetlands are valued for their water purifying characteristics (Costanza et al. 2017), this model suggests that this ecosystem service can be enhanced by harvesting. However, while N and P retention increase with harvesting, the ratio at which N and P are retained pushes the system to be more N limited, making it more sensitive to cyanobacterial blooms. On the other hand Elser et al. (2009) showed a shift towards more P limited lake systems in Europe and North America as a result of increased N deposition. For Africa N deposition has been lower, but is now increasing faster compared with Europe and North America (Dentener et al. 2006).

5.4.5 SENSITIVITY ANALYSIS

The global sensitivity analysis method that was used here (Saltelli et al. 2000) is novel in ecological modelling. An advantage over the more commonly used OAT approach was the identification of a set of parameters that relative to each other explained the outputs (Figure 5.9). This is valuable in identifying which parameters need further attention and which can be left as they are (Cariboni et al. 2007). The method also provides quality assurance by facilitating discussion on the biological or physico-chemical explanation behind the dependence of the output on the input parameters (Saltelli et al. 2000). The relatively low R^2 values for the models describing N and P retention may indicate that the OAT pre-screening did not identify the most influential parameters for retention. It is worth investigating this further by looking at other global SA methods in a future study (e.g. Makler-Pick et al. 2011).

The results of the sensitivity analysis (Figure 5.9) illustrated mostly logical (biologically meaningful) relationships and confirmed the importance of re-translocation. As expected, higher assimilation led to higher biomass, and higher mortality to lower biomass. For water quality, higher re-translocation led to higher concentrations in the water by keeping more N and P in the vegetation when it senesced and reducing uptake from the water. Similarly, retention was lower with higher re-translocation because of less N and P in aboveground dead biomass. Based on a study of macrophytes in wetlands in various regions and with different nutrient status, Rejmánková (2005) suggested that re-translocation in emergent macrophytes depends on the inorganic N and P concentrations in the water, with lower concentrations increasing re-translocation as a survival strategy. Higher mortality led to higher retention (more dead biomass) and higher fragmentation and leaching to lower retention (less accumulating dead biomass). All parameters that define papyrus growth, mortality and decay processes are also related to system conditions (e.g. temperature), which were not all modelled. Because some of the parameter values were derived from other wetland types or plant species, or calibrated or estimated (Appendix 5.3), empirical studies on papyrus systems are needed to obtain more evidence-based values for these parameters. Other influential parameters, such as inflow concentrations and porosity can be measured easily and the values used in the model were realistic. Small increases or decreases of NO_3, NH_4 and OP in the inflow led to longer or shorter periods of nutrient limitation, which highlights the impact of inflow concentrations on water quality. This impact was higher under permanent flooding conditions than with seasonal flooding. With seasonal flooding there was more nutrient limitation, higher mortality, more re-growth, a higher N and P uptake and lower concentrations in the water compared with permanent flooding. Similarly, retention was more sensitive to inflow concentrations with seasonal flooding, when longer periods of nutrient limitation led to more mortality and more accumulation of dead biomass.

5.4.6 MODEL EVALUATION AND FUTURE DEVELOPMENT

An evaluation of the quality and credibility of a model encompasses the whole process of model development, calibration, analysis and application (Augusiak et al. 2014). Based on the available data from field studies of papyrus, the numerical model used, and the results of the sensitivity analysis and comparison with literature data, it can be concluded that Papyrus Simulator provides a good representation of the main nutrient cycles in rooted papyrus wetlands and allows a better understanding of the impact of water regime and harvesting on N and P retention. If validation means "that a model is acceptable for its intended use" (Rykiel 1996), the model could be considered validated. However, validation in the sense of "model output corroboration" (Augusiak et al.2014) would require a more rigorous comparison of model output with independent time series data from different wetland sites. Wider application of Papyrus Simulator to specific wetland sites for decision-making purposes would require coupling of the model to a spatially defined hydrology model. While currently good datasets from sufficiently long time periods are not available, we hope that increasing availability of data from monitoring programmes and remote sensing will provide opportunities for further model development and a full validation.

The model may also be developed further in other directions. In the current version of the model oxygen only influences nitrification and denitrification. Other processes (e.g. fragmentation, leaching and assimilation) are also dependent on environmental factors like temperature, oxygen, pH and irradiance. For application of the model to all papyrus wetlands from the Middle East to Southern Africa, knowledge of how these factors vary with altitude and longitude and how they influence biomass, retention and water quality is crucial. For example, under the Mediterranean climate, Serag (2003) found a low AGB in winter (low temperatures and radiation) and high AGB in summer (high temperatures and radiation) and vice versa for BGB. The relationship between altitude and biomass is also complex. Temperature is lower at higher altitude (Naivasha in Kenya at 1900 m on average below 20°C compared with the Sudd wetland in South-Sudan at 400 m average 30°C), but radiation may be higher, which favours C4 photosynthesis. Higher temperature stimulates growth, but also leads to higher respiratory losses (Jones and Muthuri 1985). Higher temperature may increase decomposition and, therefore, reduce retention, but assimilation may also be higher. If this would result in higher mortality, this could increase retention. The empirical data from literature did not reveal a clear relationship between altitude, biomass and productivity

(Table 5.4), which emphasizes the need for experiments targeting the processes related to growth and mortality.

5.4.7 POTENTIAL FOR APPLICATION

The potential application of Papyrus Simulator for scientists and decision makers can be local, regional and global. For a specific papyrus wetland, the model can be linked to a site-specific hydrological model and developed into a spatially explicit tool for spatial planning or (economic) valuation of N and P retention. The model could also be applied to design constructed wetlands for wastewater treatment, to predict N and P removal rates and evaluate harvesting regimes. At a regional level, the model can improve our understanding of N and P cycling and retention and how they are affected by changes in climate or land use, identify knowledge gaps and help focus empirical research. With a more elaborated carbon section, the model can address questions on the role of papyrus wetlands in the carbon cycle and quantify their expected role as net carbon sink (Moomaw et al. 2018). A comparison with growth models of other emergent plants, e.g. *Phragmites australis* and *Typha sp.* (Asaeda et al. 2000; Tanaka et al. 2004; Asaeda et al. 2008), could identify strength and weaknesses of these respective modelling approaches and lead to mutual benefits. Combined with existing vegetation models and global hydrological and climate models, Papyrus Simulator could contribute to a global model which quantifies ecosystem services that contribute to the Sustainable Development Goals (Janse et al. 2019). Such a global model can quantify the loss of (economic) value from conversion of wetlands for food security (Conceição et al. 2016) and provide evidence of impact on human well-being.

5.4.8 CONCLUSION

In conclusion, Papyrus Simulator is suitable for studying processes related to N and P retention and the effect of harvesting in papyrus wetlands. Absolute and relative retention of P was lower than N retention and the main mechanism for both was peat accumulation. This led to a predominantly N-limited environment by reducing the N:P ratio in the water. Absence of surface water during part of the year resulted in a reduction of biomass, mainly aboveground. Harvesting increased retention from 7% to more than 40% for N, and for P from 4% to 40%. The global sensitivity analysis was successful in identifying the relative contribution of the model inputs to explaining model outputs. The most influential parameters were related to assimilation, mortality, decay, re-translocation, inflow concentrations and soil porosity. The main mechanism for retention was peat formation, which would be relatively unaffected when combined with seasonal agriculture and sustainable harvesting of papyrus, but would be completely lost when wetlands are converted to agriculture more permanently. Papyrus Simulator can contribute to a global modelling effort to quantify ecosystem services and contribute to achieving the SDGs related to food, water, climate and biodiversity.

Appendix 5.1: State variables

variable	description	#	initial value	unit	source
OPADS	OP adsorbed	1	1	g P/m^2	a
OPP	OP pore water	2	1	g P/m^2	a
OPS	OP surface water	3	1	g P/m^2	a
CAGB	C aboveground biomass	4	1853	g C/m^2	b
CBGB	C belowground biomass	5	1570	g C/m^2	b
CDAGB	C dead aboveground biomass	6	335	g C/m^2	c
CDBGB	C dead belowground biomass	7	284	g C/m^2	c
DONP	DON pore water	8	1.1	g N/m^2	e
DONS	DON surface water	9	1.1	g N/m^2	e
DOPP	DOP pore water	10	1	g P/m^2	a
DOPS	DOP surface water	11	1	g P/m^2	a
NAGB	N aboveground biomass	12	44	g N/m^2	d
NBGB	N belowground biomass	13	31	g N/m^2	d
NDAGB	N dead aboveground biomass	14	7.9	g N/m^2	c
NDBGB	N dead belowground biomass	15	5.6	g N/m^2	c
NH4P	ammonium pore water	16	0.5	g N/m^2	e
NH4S	ammonium surface water	17	0.5	g N/m^2	e
NO3P	nitrate pore water	18	0.05	g N/m^2	e
NO3S	nitrate surface water	19	0.05	g N/m^2	e
PAGB	P aboveground biomass	20	2.61	g P/m^2	f
PBGB	P belowground biomass	21	2.78	g P/m^2	f
PDAGB	P dead aboveground biomass	22	0.47	g P/m^2	c
PDBGB	P dead belowground biomass	23	0.5	g P/m^2	c
POCP	particulate organic carbon pore water	24	20	g C/m^2	a
POCS	particulate organic carbon surface water	25	20	g C/m^2	a
PONP	PON pore water	26	0.9	g N/m^2	e
PONS	PON surface water	27	0.9	g N/m^2	e
POPP	POP pore water	28	1	g P/m^2	a
POPS	POP surface water	29	1	g P/m^2	a
Water	water volume	30	0.2	m^3/m^2	a

a estimate, b average of Boar et al. (1999) Jones and Humphries (2002) Boar (2006), c calculated, see Online Resource 1, d average of Boar et al. (1999) Boar (2006), e Muthuri and Jones (1997), f Boar (2006)

Appendix 5.2: Flows (processes) in the model

process	description	formula	#
OP_adsorption	adsorption of OP	IF(OPADS<OPADS_eq)THEN((1-OPADS/OPADS_max)*(OPADS_eq-OPADS))ELSE(0)	31
OP_desorption	desorption of OP	(IF(OPADS>OPADS_eq)THEN(OPADS/OPADS_max*(OPADS-OPADS_eq))ELSE(0))	32
OP_diffusion	OP diffusion between surface and pore water	IF(surface_water>0)THEN(K_OP_diffusion*((OPS/surface_water)-(OPP/soil_depth))/((soil_depth+surface_water)/2))ELSE(0)	33
OP_inflow_lake	OP inflow lake	OP_conc_lake*lake_Inflow	34
OP_uptake	uptake by papyrus	IF(PBGB/BGB<=P_conc_BGB_lit)THEN(max_OP_uptake*(PAGB+PBGB)*(1-(PAGB+PBGB)/(P_max_AGB+P_max_BGB))*limit_OP_uptake)ELSE(0)	35
OPP_recharge	groundwater recharge	OPP_conc*recharge	36
OPP_inflow_river	OP inflow in pore water	IF(surface_water=0)THEN(OP_conc_river*river_inflow)ELSE IF(surface_water<Ksurf)AND(surface_water>0)THEN ((1-surface_water/Ksurf)*(OP_conc_river*river_inflow)) ELSE (0)	37
OPS_inflow_river	OP inflow in surface water	IF(surface_water>=Ksurf) THEN(OP_conc_river*river_inflow) ELSE IF(surface_water<Ksurf)AND(surface_water>0)THEN (surface_water/Ksurf*OP_conc_river*river_inflow) ELSE (0)	38
OPS_outflow	OP outflow	OPS_conc*outflow	39
OPSP_drain	NO_3 draining from surface to pore water	IF(surface_water=0)THEN(OPS)ELSE IF(surface_water<Ksurf)AND(surface_water>0)THEN(OPS*(1-surface_water/Ksurf))ELSE(0)	40
C_translocation	translocation	IF(CAGB>0)AND(CBGB>0)AND(C_AGB_to_BGB_optimal_ratio<(CAGB/CBGB))THEN(TLDOWNmax_C*((CAGB/CBGB)/((CAGB/CBGB)+Ktldown_C))) ElSE IF (CAGB>=0)AND(CBGB>0)AND (C_AGB_to_BGB_optimal_ratio>(CAGB/CBGB)) THEN (-TLUPmax_C*((1)/((1+Ktlup_C*CAGB/CBGB)))ELSE(0)	41
CAGB_assimilation	CO_2 uptake AGB	max_assimilation_constant*CAGB*(1-(CAGB/(C_conc_AGB_lit*K_assim))) *limit_NP_ass*limit_radiance	42
CAGB_death	death of AGB	CAGB*CAGB_death_constant	43
CAGB_harvesting	harvesting AGB	IF(harvest_batch_yes=1)AND(harvest_regular_yes=0)THEN(PULSE(harvest_%_of_AGB*CAGB, harvest_day_batch,harvest_interval_batch))ELSE IF(harvest_regular_yes=1)AND(harvest_batch_yes=0)THEN(PULSE(harvest_in_g_CAGB, harvest_day_regular,harvest_interval_regular))ELSE IF (harvest_batch_yes=1)AND(harvest_regular_yes=1)THEN(PULSE(harvest_%_of_AGB*CAGB, harvest_day_batch,harvest_interval_batch)+PULSE(harvest_in_g_CAGB,harvest_day_regular,harvest_interval_regular))ELSE(0)	44
CAGB_respiration	respiration AGB	CAGB*maint_coeff_AGB+CAGB_assimilation*growth_coeff_AGB	45
CBGB_death	death of BGB	IF(CNBGB_ratio>75)THEN(CAGB_death_constant*CBGB)ELSE(CBGB*CBGB_death_constant)	46
CBGB_respiration	respiration BGB	IF(C_translocation>0)THEN(CBGB*maint_coeff_BGB+C_translocation*growth_coeff_BGB)ELSE(CBGB*maint_coeff_BGB)	47
CDAGB_fragmentation	fragmentation of DAGB	IF(surface_water>=Ksurf)THEN(CDAGB*CDAGB_frag_constant)ELSE IF(surface_water<Ksurf)AND(surface_water>0) THEN((surface_water/Ksurf)*CDAGB*CDAGB_frag_constant)ELSE(0)	48
CDAGB_leaching	leaching from DAGB	IF(surface_water>=Ksurf)THEN(CDAGB_leach_constant*CAGB_death) ELSE IF(surface_water<Ksurf)AND(surface_water>0)THEN ((surface_water/Ksurf)*CDAGB_leach_constant*CAGB_death)ELSE(0)	49
CDBGB_fragmentation	fragmentation of DBGB	IF(pore_water>=Kpor)THEN(CDBGB*CDBGB_fragmentation_constant) ELSE IF(pore_water<Kpor)AND(pore_water>0)THEN ((pore_water/Kpor)*CDBGB*CDBGB_fragmentation_constant)ELSE(0)	50
CDBGB_leaching	leaching from DBGB	IF(pore_water>=Kpor)THEN(CDBGB_leach_constant*CBGB_death)ELSE IF(pore_water<Kpor)AND(pore_water>0)THEN((pore_water/Kpor)*CDBGB_leach_constant*CBGB_death)ELSE(0)	51
denitrification_P	denitrification	IF(pore_water>=Kpor)THEN(K_denitri*NO3P*(1-mode))ELSE IF(pore_water<Kpor)AND(pore_water>0) THEN((pore_water/Kpor)*(K_denitri*NO3P*(1-mode)))ELSE(0)	52
DON_diffusion	DON diffusion between	IF(surface_water>0) THEN(K_DON_diffusion*((DONS/surface_water)-(DONP/soil_depth))/((surface_water+soil_depth)/2)) ELSE(0)	53

	surface and pore water		
DON_inflow_lake	DON inflow lake	lake_Inflow*DON_conc_lake	54
DONP_inflow_river	DON inflow in pore water	IF(surface_water=0)THEN(DON_conc_river*river_inflow)ELSE IF(surface_water<Ksurf)AND(surface_water>0)THEN ((1-surface_water/Ksurf)*(DON_conc_river*river_inflow)) ELSE (0)	55
DONP_mineral	mineralisation	IF(pore_water>=Kpor)THEN(DON_mineral_const*DONP)ELSE IF(pore_water<Kpor)AND(pore_water>0) THEN((pore_water/Kpor)*DON_mineral_const*DONP)ELSE(0)	56
DONP_recharge	groundwater recharge	DONP_conc*recharge	57
DONS_inflow_river	DON inflow in surface water	IF(surface_water>=Ksurf) THEN(DON_conc_river*river_inflow) ELSE IF(surface_water<Ksurf)AND(surface_water>0)THEN (surface_water/Ksurf*DON_conc_river*river_inflow) ELSE (0)	58
DONS_mineral	mineralisation in surface water	IF(surface_water>=Ksurf)THEN(DON_mineral_const*DONS)ELSE IF(surface_water<Ksurf)AND(surface_water>0) THEN((surface_water/Ksurf)*DON_mineral_const*DONS)ELSE(0)	59
DONS_outflow	outflow of DON	outflow*DONS_conc	60
DONSP_drain	DON draining from surface to pore water	IF(surface_water=0)THEN(DONS)ELSE IF(surface_water<Ksurf) AND(surface_water>0)THEN(DONS*(1-surface_water/Ksurf))ELSE(0)	61
DOP_diffusion	DOP diffusion between surface and pore water	IF(surface_water>0)THEN(K_DOP_diffusion*((DOPS/surface_water)- (DOPP/soil_depth))/((surface_water+soil_depth)/2))ELSE(0)	62
DOP_inflow_lake	DOP inflow lake	DOP_conc_lake*lake_Inflow	63
DOPP_mineral	mineralisation	IF(pore_water>=Kpor)THEN(DOP_mineral_const*DOPP)ELSE IF(pore_water<Kpor)AND(pore_water>0)THEN((pore_water/Kpor)*DOP _mineral_const*DOPP)ELSE(0)	64
DOPP_recharge	groundwater recharge	DOPP_conc*recharge	65
DOPP_inflow_river	DOP inflow in pore water	IF(surface_water=0)THEN(DOP_conc_river*river_inflow)ELSE IF(surface_water<Ksurf)AND(surface_water>0)THEN ((1-surface_water/Ksurf)*(DOP_conc_river*river_inflow)) ELSE (0)	66
DOPS_inflow_river	DOP inflow in surface water	IF(surface_water>=Ksurf) THEN(DOP_conc_river*river_inflow) ELSE IF(surface_water<Ksurf)AND(surface_water>0)THEN (surface_water/Ksurf*DOP_conc_river*river_inflow) ELSE (0)	67
DOPS_mineral	mineralisation in surface water	IF(surface_water>=Ksurf)THEN(DOP_mineral_const*DOPS)ELSE IF(surface_water<Ksurf)AND(surface_water>0)THEN((surface_water/Ks urf)*DOP_mineral_const*DOPS)ELSE(0)	68
DOPS_outflow	outflow of DOP	DOPS_conc*outflow	69
DOPSP_drain	DOP draining from surface to pore water	IF(surface_water=0)THEN(DOPS)ELSEIF(surface_water<Ksurf)AND(surfac e_water>0)THEN(DOPS*(1-surface_water/Ksurf))ELSE(0)	70
evaporation	evaporation	evaporation_rate*0.001	71
lake_inflow	lake inflow	IF(surface_water<max_depth)THEN(lake_inflow_rate)ELSE(0)* wet_yes_or_no	72
N_retrans	retranslocation	IF(CNAGB_ratio>0)THEN(CAGB_death*N_retrans_constant/CNAGB_rati o)ELSE(0)	73
N_translocation	translocation	IF(NAGB>=0)AND(NBGB>0)AND(N_AGB_to_BGB__optimal_ratio< (NAGB/NBGB))AND(N_conc_BGB<=N_conc_BGB_lit)THEN (tldownmax_N*((NAGB/NBGB)/((NAGB/NBGB)+Ktldown_N)))ElSE IF (NAGB>=0)AND(NBGB>0)AND(N_AGB_to_BGB__optimal_ratio>(NAGB/ NBGB))AND(N_conc_AGB<=N_conc_AGB_lit)THEN(-tlupmax_N* (1/(1+Ktlup_N*(NAGB/NBGB))))ELSE(0)	74
NAGB_death	death of AGB	IF(CNAGB_ratio>0)THEN(CAGB_death*(1- N_retrans_constant)/CNAGB_ratio)ELSE(0)	75
NAGB_harvesting	harvesting AGB	IF(CNAGB_ratio>0)THEN(CAGB_harvesting/CNAGB_ratio)ELSE(0)	76
NBGB_death	death of BGB	IF(CNBGB_ratio>0) THEN(CBGB_death/CNBGB_ratio) ELSE (0)	77
NDAGB_fragmentation	fragmentation of DAGB	IF(surface_water>=Ksurf)THEN(CDAGB_fragmentation/CNDAGB_ratio) ELSE IF(surface_water<Ksurf)AND(surface_water>0)THEN ((surface_water/Ksurf)*CDAGB_fragmentation/CNDAGB_ratio)ELSE(0)	78
NDAGB_leaching	leaching from DABG	IF(surface_water>=Ksurf)THEN(CDAGB_leaching/CNDAGB_ratio)ELSE IF(surface_water<Ksurf)AND(surface_water>0) THEN((surface_water/Ksurf)*CDAGB_leaching/CNDAGB_ratio)ELSE(0)	79

NDBGB_fragmentation	fragmentation of DBGB	IF(pore_water>=Kpor)THEN(CDBGB_fragmentation/CNDBGB_ratio)ELSE IF(pore_water<Kpor)AND(pore_water>0) THEN ((pore_water/Kpor)*CDBGB_fragmentation/CNDBGB_ratio)ELSE(0)	80
NDBGB_leaching	leaching from DBGB	IF(pore_water>=Kpor)THEN(CDBGB_leaching/CNDBGB_ratio)ELSE IF(pore_water<Kpor)AND(pore_water>0) THEN ((pore_water/Kpor)*CDBGB_leaching/CNDBGB_ratio)ELSE(0)	81
NH4_diffusion	NH_4 diffusion between surface and pore water	IF(surface_water>0) THEN(K_NH4_diffusion*((NH4S/surface_water)-(NH4P/soil_depth))/((surface_water+soil_depth)/2)) ELSE(0)	82
NH4_inflow_lake	NH_4 inflow lake	lake_Inflow*NH4_conc_lake	83
NH4_uptake	uptake by papyrus	IF(NBGB/BGB<=N_conc_BGB_lit)THEN(max_NH4_uptake*(NAGB+NBGB)*(1-(NAGB+NBGB)/(N_max_AGB+N_max_BGB))*limit_NH4_uptake) ELSE(0)	84
NH4P_inflow_river	NH_4 inflow in pore water	IF(surface_water=0)THEN(NH4_conc_river*river_inflow)ELSE IF(surface_water<Ksurf)AND(surface_water>0)THEN ((1-surface_water/Ksurf)*(NH4_conc_river*river_inflow)) ELSE (0)	85
NH4P_recharge	groundwater recharge	NH4P_conc*recharge	86
NH4S_inflow_river	NH_4 inflow in surface water	IF(surface_water>=Ksurf) THEN(NH4_conc_river*river_inflow) ELSE IF(surface_water<Ksurf)AND(surface_water>0)THEN (surface_water/Ksurf*NH4_conc_river*river_inflow) ELSE (0)	87
NH4S_outflow	NH_4 outflow	outflow*NH4S_conc	88
NH4SP_drain	NH_4 draining from surface to pore water	IF(surface_water=0)THEN(NH4S)ELSE IF(surface_water<Ksurf) AND(surface_water>0)THEN(NH4S*(1-surface_water/Ksurf))ELSE(0)	89
nitrification_P	nitrification in pore water	IF(pore_water>=Kpor)THEN(K_nitri*mode*NH4P)ELSE IF(pore_water<Kpor)AND(pore_water>0) THEN((pore_water/Kpor)*(K_nitri*mode*NH4P))ELSE(0)	90
nitrification_S	nitrification in surface water	IF(surface_water>=Ksurf)THEN(K_nitri*NH4S)ELSE IF(surface_water<Ksurf)AND(surface_water>0) THEN((surface_water/Ksurf)*(K_nitri*NH4S))ELSE (0)	91
NO3_diffusion	NO_3 diffusion between surface and pore water	IF(surface_water>0) THEN(K_NO3_diffusion*((NO3SP/surface_water)-(NO3PP/soil_depth_P))/((surface_water+soil_depth_P)/2))ELSE(0)	92
NO3_inflow_lake	NO_3 inflow lake	lake_Inflow*NO3_conc_lake	93
NO3_uptake	uptake by papyrus	IF(NBGB/BGB<=N_conc_BGB_lit)THEN(max_NO3_uptake*(NAGB+NBGB)*(1-(NAGB+NBGB)/(N_max_AGB+N_max_BGB))*limit_NO3_uptake) ELSE(0)	94
NO3P_inflow_river	NO_3 inflow in pore water	IF(surface_water=0)THEN(NO3_conc_river*river_inflow)ELSE IF(surface_water<Ksurf)AND(surface_water>0)THEN ((1-surface_water/Ksurf)*(NO3_conc_river*river_inflow)) ELSE (0)	95
NO3P_recharge	groundwater recharge	NO3P_conc*recharge	96
NO3S_inflow_river	NO_3 inflow in surface water	IF(surface_water>=Ksurf) THEN(NO3_conc_river*river_inflow) ELSE IF(surface_water<Ksurf)AND(surface_water>0)THEN (surface_water/Ksurf*NO3_conc_river*river_inflow) ELSE (0)	97
NO3S_outflow	NO_3 outflow	outflow*NO3S_conc	98
NO3SP_drain	NO_3 draining from surface to pore water	IF(surface_water=0)THEN(NO3S)ELSE IF(surface_water<Ksurf) AND(surface_water>0)THEN(NO3S*(1-surface_water/Ksurf))ELSE(0)	99
outflow	outflow to lake	IF(surface_water>max_depth)THEN(river_inflow+precipitation-evaporation-recharge)ELSE(0)	100
P_retrans	retranslocation	IF(CPAGB_ratio>0)THEN(P_retrans_constant*CAGB_death/CPAGB_ratio)ELSE(0)	101
P_translocation	translocation	IF(PAGB>=0)AND(PBGB>0)AND(P_AGB_to_BGB__optimal_ratio< (PAGB/PBGB))AND(P_conc_BGB<=P_conc_BGB_lit)THEN (tldownmax_P*((PAGB/PBGB)/((PAGB/PBGB)+Ktldown_P)))ElSE IF (PAGB>=0)AND (PBGB>0)AND(P_AGB_to_BGB__optimal_ratio> (PAGB/PBGB))AND (P_conc_AGB<=P_conc_AGB_lit)THEN(-tlupmax_P* (1/(1+Ktlup_P*(PAGB/PBGB))))ELSE(0)	102
PAGB_death	death of AGB	IF(CPAGB_ratio>0)THEN(CAGB_death*(1-P_retrans_constant)/ CPAGB_ratio)ELSE(0)	103
PAGB_harvesting	harvesting AGB	IF(CPAGB_ratio>0)THEN(CAGB_harvesting/CPAGB_ratio)ELSE(0)	104

Name	Description	Equation	#
PBGB_death	death of BGB	IF(CPBGB_ratio>0) THEN(CBGB_death/CPBGB_ratio) ELSE (0)	105
PDAGB_fragmentation	fragmentation of DAGB	IF(surface_water>=Ksurf)THEN(CDAGB_fragmentation/CPDAGB_ratio)ELSE IF(surface_water<Ksurf)AND(surface_water>0)THEN ((surface_water/Ksurf)*CDAGB_fragmentation/CPDAGB_ratio)ELSE(0)	106
PDAGB_leaching	leaching from DABG	IF(surface_water>=Ksurf)THEN(CDAGB_leaching/CPDAGB_ratio)ELSE IF(surface_water<Ksurf)AND(surface_water>0)THEN((surface_water/ Ksurf) *CDAGB_leaching/CPDAGB_ratio)ELSE(0)	107
PDBGB_fragmentation	fragmentation of DBGB	IF(pore_water>=Kpor)THEN(CDBGB_fragmentation/CPDBGB_ratio)ELSE IF(pore_water<Kpor)AND(pore_water>0) THEN((pore_water/Kpor)* CDBGB_fragmentation/CPDBGB_ratio)ELSE(0)	108
PDBGB_leaching	leaching from DBGB	IF(pore_water>=Kpor)THEN(CDBGB_leaching/CPDBGB_ratio)ELSE IF(pore_water<Kpor)AND(pore_water>0)THEN((pore_water/Kpor)* CDBGB_leaching/CPDBGB_ratio)ELSE(0)	109
POC_drain	POC draining from surface to pore water	IF(surface_water=0)THEN(POCS)ELSE IF(surface_water<Ksurf)AND(surface_water>0)THEN(POCS*(1-surface_water/Ksurf))ELSE(0)	110
POC_settling	settling of POC	POC_settling_rate*POCS	111
POCP_hydrolysis	hydrolysis of POC in pore water	IF(pore_water>=Kpor)THEN(POCP*POC_hydrolysis_constant)ELSE IF(pore_water<Kpor)AND(pore_water>0)THEN((pore_water/Kpor)*POCP*POC_hydrolysis_constant)ELSE(0)	112
POCS_hydrolysis	hydrolysis of POC in surface water	IF(surface_water>=Ksurf)THEN(POCS*POC_hydrolysis_constant)ELSE IF(surface_water<Ksurf)AND(surface_water>0) THEN((surface_water/Ksurf)*POCS*POC_hydrolysis_constant)ELSE(0)	113
PON_inflow_lake	PON inflow lake	lake_Inflow*PON_conc_lake	114
PON_settling	settling of PON	PONS*PON_settling_constant	115
PONP_hydrolysis	hydrolysis of PON in pore water	IF(pore_water>=Kpor)THEN(PONP*PON_hydrolysis_const)ELSE IF(pore_water<Kpor)AND(pore_water>0) THEN((pore_water/Kpor)*PONP*PON_hydrolysis_const)ELSE(0)	116
PONP_inflow_river	PON inflow in pore water	IF(surface_water=0)THEN(PON_conc_river*river_inflow)ELSE IF(surface_water<Ksurf)AND(surface_water>0)THEN ((1-surface_water/Ksurf)*(PON_conc_river*river_inflow)) ELSE (0)	117
PONP_recharge	groundwater recharge	PONP_conc*recharge	118
PONS_hydrolysis	hydrolysis of PON in surface water	IF(surface_water>=Ksurf)THEN(PONS*PON_hydrolysis_const)ELSE IF(surface_water<Ksurf)AND(surface_water>0) THEN((surface_water/Ksurf)*PONS*PON_hydrolysis_const)ELSE(0)	119
PONS_inflow_river	PON inflow in surface water	IF(surface_water>=Ksurf) THEN(PON_conc_river*river_inflow) ELSE IF(surface_water<Ksurf)AND(surface_water>0)THEN (surface_water/Ksurf*PON_conc_river*river_inflow) ELSE (0)	120
PONS_outflow	PON outflow	outflow*PONS_conc	121
PONSP_drain	PON draining from surface to pore water	IF(surface_water=0)THEN(PONS)ELSE IF(surface_water<Ksurf) AND(surface_water>0)THEN(PONS*(1-surface_water/Ksurf))ELSE(0)	122
POP_inflow_lake	POP inflow lake	lake_Inflow*POP_conc_lake	123
POP_settling	settling of POP	POPS*POP_settling_constant	124
POPP_hydrolysis	hydrolysis of POP in pore water	IF(pore_water>=Kpor)THEN(POPP*POP_hydrolysis_constant)ELSE IF(pore_water<Kpor)AND(pore_water>0) THEN((pore_water/Kpor)*POPP*POP_hydrolysis_constant)ELSE(0)	125
POPP_recharge	groundwater recharge	POPP_conc*recharge	126
POPP_inflow_river	POP inflow in pore water	IF(surface_water=0)THEN(POP_conc_river*river_inflow)ELSE IF(surface_water<Ksurf)AND(surface_water>0)THEN ((1-surface_water/Ksurf)*(POP_conc_river*river_inflow)) ELSE (0)	127
POPS_hydrolysis	hydrolysis of POP in surface water	IF(surface_water>=Ksurf)THEN(POPS*POP_hydrolysis_constant)ELSE IF(surface_water<Ksurf)AND(surface_water>0) THEN ((surface_water/Ksurf)*POPS*POP_hydrolysis_constant)ELSE(0)	128
POPS_inflow_river	POP inflow in surface water	IF(surface_water>=Ksurf) THEN(POP_conc_river*river_inflow) ELSE IF(surface_water<Ksurf)AND(surface_water>0)THEN (surface_water/Ksurf*POP_conc_river*river_inflow) ELSE (0)	129
POPS_outflow	POP outflow	outflow*POPS_conc	130
POPSP_drain	POP draining from surface to pore water	IF(surface_water=0)THEN(POPS)ELSE IF(surface_water<Ksurf)AND (surface_water>0)THEN(POPS*(1-surface_water/Ksurf))ELSE(0)	131
precipitation	precipitation	rainfall_rate*0.001	132

| recharge | groundwater recharge | fraction_out*porewater_free | 133 |
| river_inflow | inflow from catchment | river_inflow_rate | 134 |

Appendix 5.3: Variables and constants of the model

Name	description	formula or value	unit	source
AGB	aboveground biomass	CAGB/C_conc_AGB_lit	g DW/m^2	
AGB_total_biomass_ratio	fraction of AGB DW of total plant DW	AGBlitavg/ papyrus_biomass_litavg	-	
AGBlitavg	average value from literature for aboveground biomass	3489	g DW/m^2	average of Muthuri *et al.*, 1989; Jones and Muthuri, 1997 and Boar, 2006
OP_conc_lake	OP concentration in lake	0.3	g P/m^3	estimate
OP_conc_river	OP concentration in inflow	0.5	g P/m^3	estimate
OPADS_eq	the equilibrium of OP adsorbed per m^2 at a certain OP concentration	OPADS_max*OPP_conc/ (OPP_conc+km_ads)	g P/m^2	
OPADS_max	the maximum of OP that can be adsorbed per m^2	OPADS_maxdw*bulkdens* soil_depth	g P/m^2	
OPADS_maxdw	adsorption capacity per gram dw of the soil	0.004	g P/g DW	Kelderman *et al.*, 2007
OPP_conc	OP concentration in pore water	IF(pore_water>0)THEN(OPP/ pore_water)ELSE(0)	g P/m^3	
OPS_conc	OP concentration in surface water	IF(surface_water>0)THEN(OPS/ surface_water)ELSE(0)	g P/m^3	
BGB	belowground biomass	CBGB/C_conc_BGB_lit	g DW/m^2	
BGB_total_biomass_ratio	fraction of BGB DW of total plant DW	BGBlitavg/ papyrus_biomass_litavg	-	
BGBlitavg	average value from literature for belowground biomass	3928	g DW/m^2	average of Jones and Muthuri, 1997 and Boar, 2006
bulkdens	bulkdensity of the soil	180000	g DW/m^3	Jones and Muthuri, 1997
C_AGB_to_BGB_optimal_ratio	optimal C AGB to C BGB ratio	CAGBlitavg/CBGBlitavg	-	
C_conc_AGB	carbon in AGB	IF(AGB>0)THEN(CAGB/AGB) ELSE(0)	g C/g DW	
C_conc_AGB_lit	carbon in AGB calculated on literature average	CAGBlitavg/AGBlitavg	g C/g DW	
C_conc_BGB	carbon in BGB	IF (BGB>0) THEN(CBGB/BGB) ELSE(0)	g C/g DW	
C_conc_BGB_lit	carbon in BGB calculated on literature average	CBGBlitavg/BGBlitavg	g C/g DW	
C_conc_papyrus	carbon in papyrus	IF(papyrus_biomass>0)THEN ((CAGB+CBGB)/ papyrus_biomass)ELSE(0)	g C/g DW	
C_max_AGB	maximum amount of carbon in AGB	papyrus_max_biomass* C_conc_AGB_lit* AGB_total_biomass_ratio	g C/m^2	
C_max_BGB	maximum amount of carbon in AGB	papyrus_max_biomass* C_conc_BGB_lit* BGB_total_biomass_ratio	g C/m^2	
C_papyrus	amount of carbon in papyrus	CAGB+CBGB	g C/m^2	
CAGB_death_constant	death rate of above ground biomass	0.0057	day^{-1}	calibrated
CAGBlitavg	average value from literature for C in AGB	1853	g C/m^2	average of Boar *et al.*, 1999 Jones and Humphries, 2002 and Boar, 2006
CBGB_death_constant	death rate of below ground biomass	0.0014	day^{-1}	van der Peijl and Verhoeven, 1999
CBGBlitavg	average value from literature for C in BGB	1570	g C/m^2	average of Boar *et al.*, 1999 Jones and Humphries, 2002 and Boar, 2006

CDABG_leach_constant	maximum fraction leached of above ground biomass	0.432	-	van der Peijl and Verhoeven, 1999
CDAGB_frag_constant	fragmentation rate of above ground biomass	0.0005	day^{-1}	estimated
CDBGB_frag_constant	fragmentation rate of below ground biomass	$8.34*10^{-4}$	day^{-1}	van der Peijl and Verhoeven, 1999
CDBGB_leach_constant	maximum fraction leached of below ground biomass	0.486	-	van der Peijl and Verhoeven, 1999
CNAGB_ratio	C to N ratio in above ground biomass	IF (NAGB>0) THEN(CAGB/NAGB) ELSE(0)	g C/g N	
CNBGB_ratio	C to N ratio in below ground biomass	IF(NBGB>0)THEN(CBGB/ NBGB)ELSE(0)	g C/g N	
CNDAGB_ratio	C to N ratio in dead above ground biomass	CDAGB/NDAGB	g C/g N	
CNDBGB_ratio	C to N ratio in dead below ground biomass	CDBGB/NDBGB	g C/g N	
CPAGB_ratio	C to P ratio in above ground biomass	IF(PAGB>0) THEN (CAGB/PAGB) ELSE(0)	g C/g P	
CPBGB_ratio	C to P ratio in below ground biomass	IF(PBGB>0)THEN (CBGB/PBGB)ELSE(0)	g C/g P	
CPDAGB_ratio	C to P ratio in dead above ground biomass	CDAGB/PDAGB	g C/g P	
CPDBGB_ratio	C to P ratio in dead below ground biomass	CDBGB/PDBGB	g C/g P	
DON_conc_lake	DON concentration in lake	0.5	g N/m^3	estimate
DON_conc_river	DON concentration in inflow	1	g N/m^3	estimate
DON_mineral_const	mineralization constant	0.0002	day^{-1}	van Dam *et al.*, 2007
DONP_conc	concentration of dissolved organic nitrogen in pore water	IF(pore_water>0)THEN(DONP/pore_water)ELSE(0)	g N/m^3	
DONS_conc	concentration of dissolved organic nitrogen in surface water	IF(surface_water>0)THEN(DONS/surface_water)ELSE(0)	g N/m^3	
DOP_conc_lake	DOP concentration in lake	0.1	g P/m^3	estimate
DOP_conc_river	DOP concentration in inflow	0.1	g P/m^3	estimate
DOP_mineral_const	mineralization constant	0.0002	day^{-1}	estimate
DOPP_conc	concentration of dissolved organic phosphorus in pore water	IF(pore_water>0)THEN(DOPP/pore_water)ELSE(0)	g P/m^3	
DOPS_conc	concentration of dissolved organic phosphorus in surface water	IF(surface_water>0)THEN(DOPS/surface_water)ELSE(0)	g P/m^3	
evaporation_rate	evaporation rate in Naivasha region	COUNTER(0,365)	mm/day	Gaudet, 1978
fraction_out	fraction of water recharged to groundwater	0.01	day^{-1}	estimate
growth_coeff_AGB	coefficient of respiration represented by growth component, proportional to photosynthesis for AGB	0.3	day^{-1}	Bachelet *et al.*, 1989
growth_coeff_BGB	coefficient of respiration represented by growth component, proportional to photosynthesis for BGB	0.2	day^{-1}	Bachelet *et al.*, 1989

harvest_%_of_AGB	fraction of AGB harvested	value between 0 and 1	-	
harvest_batch_yes	switch to activate batch harvest	0 or 1	-	
harvest_day_batch	day of the year when first batch harvest takes place	value between 1 and 365	-	
harvest_day_regular	day of the year when first regular harvest takes place	value between 1 and 365	-	
harvest_in_g_AGB	amount harvested in grams of AGB	user defined	g DW/m^2	
harvest_in_g_CAGB	amount harvested in grams of CAGB	harvest_in_g_AGB* C_conc_AGB_lit	g C/m^2	
harvest_interval_batch	period between each batch harvest	value between 1 and 365	-	
harvest_interval_regular	period between each regular harvest	value between 1 and 365	-	
harvest_regular_yes	switch to activate regular harvest	0 or 1	-	
K_OP	half saturation constant phosphorus uptake	0.1	g P/m^3	estimate
K_OP_diffusion	diffusion rate constant for phosphorus	0.1	m^2/day	estimate
K_assim	K assimilation	5150	g DW/m^2	calibrated
K_denitri	denitrification rate	0.01	day^{-1}	estimate
K_DON_diffusion	diffusion rate constant for dissolved organic nitrogen	0.1	m^2/day	estimate
K_DOP_diffusion	diffusion rate constant for dissolved organic phosphorus	0.1	m^2/day	estimate
K_NH4	half saturation constant	0.7	g N/m^3	van Dam *et al.*, 2007
K_NH4_diffusion	diffusion rate constant for ammonium	0.1	m^2/day	estimate
K_nitri	nitrification rate	0.005	day^{-1}	estimate
K_NO3	half saturation constant	0.1	g N/m^3	van Dam *et al.*, 2007
K_NO3_diffusion	diffusion rate constant for nitrate	0.1	m^2/day	estimate
K_radiance	half saturation constant	1	MJ/m^2*day	estimate
km_ads	P concentration at which half of adsorption capacity of the soil is used	18.54	g P/m^3	van der Peijl and Verhoeven, 1999
kmN	concentration of N in the plant at which limiting factor is 0.5	N_conc_papyrus_min+ ((N_conc_papyrus_optimal- N_conc_papyrus_min)/9)	g N/g DW	
KmP	concentration of P in the plant at which limiting factor is 0.5	P_conc_papyrus_min+ ((P_conc_papyrus_optimal- P_conc_papyrus_min)/9)	g P/g DW	
Kpor	pore water constant slowing down processes at low water quantity	0.1	m^3/m^2	calibrated
Ksurf	surface water constant slowing down processes at low water quantity	0.1	m^3/m^2	calibrated
Ktldown_C	translocation constant for carbon from AGB to BGB	0.2	-	calibrated
Ktldown_N	translocation constant for nitrogen from AGB to BGB	0.2	-	calibrated
Ktldown_P	translocation constant for phosphorus from AGB to BGB	0.2	-	calibrated

Ktlup_C	translocation constant for carbon from BGB to AGB	0.2	-	calibrated
Ktlup_N	translocation constant for nitrogen from BGB to AGB	0.5	-	calibrated
Ktlup_P	translocation constant for phosphorus from BGB to AGB	0.5	-	calibrated
lake_inflow_rate	lake inflow rate in P wetland	0.12	$m^3/m^2 \ast day$	estimate
limit_OP_uptake	limitation factor for uptake of OP	OPP_conc/(OPP_conc+K_OP)	-	
limit_N_ass	N limiting factor for carbon assimilation	IF((N_conc_papyrus-N_conc_papyrus_min)/((kmN-N_conc_papyrus_min)+ (N_conc_papyrus-N_conc_papyrus_min))<(0.9-N_conc_papyrus))AND((0.9-N_conc_papyrus)>N_conc_papyrus_min)AND(N_conc_papyrus>N_conc_papyrus_min)THEN ((N_conc_papyrus-N_conc_papyrus_min)/((kmN-N_conc_papyrus_min)+ (N_conc_papyrus-N_conc_papyrus_min)))ELSE IF((N_conc_papyrus-N_conc_papyrus_min)/((kmN-N_conc_papyrus_min)+ (N_conc_papyrus-N_conc_papyrus_min))>=(0.9-N_conc_papyrus))AND((0.9-N_conc_papyrus)>N_conc_papyrus_min)AND(N_conc_papyrus>N_conc_papyrus_min) THEN(0.9)ELSE(0)	-	van der Peijl and Verhoeven, 1999
limit_NH4_uptake	limitation factor for uptake of ammonium	NH4P_conc/(NH4P_conc+ K_NH4)	-	
limit_NO3_uptake	limitation factor for uptake of nitrate	NO3P_conc/(NO3P_conc+ K_NO3)	-	
limit_NP_ass	combined N and P limiting factor for carbon assimilation	(limit_N_ass*limit_P_ass)/0.81	-	
limit_P_ass	P limiting factor for carbon assimilation	IF((P_conc_papyrus-P_conc_papyrus_min)/((kmP-P_conc_papyrus_min)+ (P_conc_papyrus-P_conc_papyrus_min))<(0.9-P_conc_papyrus))AND((0.9-P_conc_papyrus)>P_conc_papyrus_min)AND(P_conc_papyrus>P_conc_papyrus_min) THEN(P_conc_papyrus-P_conc_papyrus_min)/((kmP-P_conc_papyrus_min)+ (P_conc_papyrus-P_conc_papyrus_min))ELSE IF((P_conc_papyrus-P_conc_papyrus_min)/((kmP-P_conc_papyrus_min)+ (P_conc_papyrus-P_conc_papyrus_min))>=(0.9-P_conc_papyrus))AND((0.9-P_conc_papyrus)>P_conc_papy	-	van der Peijl and Verhoeven, 1999

		rus_min)AND(P_conc_papyrus> P_conc_papyrus_min) THEN(0.9)ELSE(0)		
limit_radiance	radiance limitation of carbon assimilation	radiance/(radiance+K_radiance)	-	
maint_coeff_AGB	maintenance coefficient for AGB	0.02	day^{-1}	Bachelet *et al.*, 1989
maint_coeff_BGB	maintenance coefficient for BGB	0.002	day^{-1}	Bachelet *et al.*, 1989
max_AGB_biomass	maximum AGB biomass	papyrus_max_biomass*AGB_total_biomass_ratio	g/m^2	
max_OP_uptake	maximum uptake of OP	0.1	day^{-1}	estimate
max_assimilation_constant	maximum assimilation of carbon by papyrus	0.17	day^{-1}	estimate
max_depth	treshold surface water depth	0.5	m^3/m^2	calibrated
max_NH4_uptake	maximum uptake of ammonium	0.1	day^{-1}	estimate
max_NO3_uptake	maximum uptake of nitrate	0.1	day^{-1}	estimate
mode	controlling factor for oxygen availability	IF(water_filled_porosity>1) OR(water_filled_porosity=1) THEN(0) ELSE IF(water_filled_porosity>water_filled_porosity_field_capacity) AND(water_filled_porosity<1) THEN((1-ater_filled_porosity)/(1-water_filled_porosity_field_capacity)) ELSE(1)	-	
N_AGB_to_BGB_optimal_ratio	optimal N AGB to N BGB ratio	NAGBlitavg/NBGBlitavg	-	
N_conc_AGB	N concentration in AGB	If (AGB>0) THEN(NAGB/AGB) ELSE (0)	g N/g DW	
N_conc_AGB_lit	nitrogen in AGB calculated on literature average	NAGBlitavg/AGBlitavg	g N/g DW	
N_conc_BGB	N concentration in BGB	IF(BGB>0)THEN(NBGB/BGB) ELSE(0)	g N/g DW	
N_conc_BGB_lit	nitrogen in BGB calculated on literature average	NBGBlitavg/BGBlitavg	g N/g DW	
N_conc_papyrus	N concentration in papyrus	IF(C_papyrus>0)THEN ((N_papyrus/C_papyrus)* C_conc_papyrus)ELSE(0)	g N/g DW	
N_conc_papyrus_min	minimum N conc in plant required for growth	0.0016	g N/g DW	van der Peijl and Verhoeven, 1999
N_conc_papyrus_optimal	optimal N conc in plant for growth	N_max_papyrus/ papyrus_max_biomass	g N/g DW	
N_max_AGB	maximum N in AGB	papyrus_max_biomass* N_conc_AGB_lit* AGB__total_biomass_ratio	g N/m^2	
N_max_BGB	maximum N in BGB	papyrus_max_biomass* N_conc_BGB_lit* BGB__total_biomass_ratio	g N/m^2	
N_max_papyrus	maximum N in papyrus biomass	N_max_AGB+N_max_BGB	g N/m^2	
N_papyrus	N in papyrus biomass	NAGB+NBGB	g N/m^2	
N_retrans_constant	fraction of nitrogen retranslocated after dying shoot	0.7	-	calibrated
NAGBlitavg	average value from literature for N in AGB	44	g N/m^2	average of Boar *et al.*, 1999 Boar, 2006

NBGBlitavg	average value from literature for N in BGB	31	g N/m^2	Boar *et al.,* 1999 Boar, 2006
NH4_conc_lake	NH4 concentration in lake	1	g N/m^3	estimate
NH4_conc_river	NH4 concentration in inflow	3	g N/m^3	estimate
NH4P_conc	concentration of NH4 in pore water	IF(pore_water>0)THEN (NH4P/pore_water)ELSE(0)	g N/m^3	
NH4S_conc	concentration of NH4 in surface water	IF(surface_water>0)THEN (NH4S/surface_water)ELSE(0)	g N/m^3	
NO3_conc_lake	NO3 concentration in lake	1	g N/m^3	estimate
NO3_conc_river	NO3 concentration in inflow	2	g N/m^3	estimate
NO3P_conc	concentration of NO3 in pore water	IF(pore_water>0)THEN (NO3P/pore_water)ELSE(0)	g N/m^3	
NO3S_conc	concentration of NO3 in surface water	IF(surface_water>0)THEN (NO3S/surface_water)ELSE(0)	g N/m^3	
P_AGB_to_BGB_optimal_ratio	optimal P AGB to P BGB ratio	PAGBlitavg/PBGBlitavg	-	
P_conc_AGB	P concentration in AGB	IF(AGB>0)THEN(PAGB/AGB) ELSE(0)	g P/g DW	
P_conc_AGB_lit	P in AGB calculated on literature average	PAGBlitavg/AGBlitavg	g P/g DW	
P_conc_BGB	P concentration in BGB	IF(BGB>0)THEN(PBGB/BGB) ELSE(0)	g P/g DW	
P_conc_BGB_lit	P in BGB calculated on literature average	PBGBlitavg/BGBlitavg	g P/g DW	
P_conc_papyrus	P concentration in papyrus	IF(C_papyrus>0)THEN ((P_papyrus/C_papyrus)* C_conc_papyrus)ELSE(0)	g P/g DW	
P_conc_papyrus_min	minimum P conc in plant required for growth	8*10^{-5}	g P/g DW	van der Peijl and Verhoeven, 1999
P_conc_papyrus_optimal	optimal P conc in plant for growth	P_max_papyrus/ papyrus_max_biomass	g P/g DW	
P_max_AGB	maximum P in AGB	papyrus_max_biomass* P_conc_AGB_lit* AGB__total_biomass_ratio	g P/m^2	
P_max_BGB	maximum P in BGB	papyrus_max_biomass* P_conc_BGB_lit* BGB__total_biomass_ratio	g P/m^2	
P_max_papyrus	maximum P in papyrus biomass	P_max_AGB+P_max_BGB	g P/m^2	
P_papyrus	P in papyrus biomass	PAGB+PBGB	g P/m^2	
P_retrans_constant	fraction of P retranslocated after dying shoot	0.77	-	van der Peijl and Verhoeven, 1999
PAGBlitavg	average value from literature for P in AGB	2.61	g P/m^2	Boar, 2006
papyrus_biomass	total amount of biomass	BGB+AGB	g DW/m^2	
papyrus_biomass_litavg	total biomass from literature	AGBlitavg+BGBlitavg	g DW/m^2	
papyrus_max_biomass	maximum papyrus biomass	8118	g DW/m^2	Muthuri *et al.,* 1989 and Jones and Muthuri, 1997
PBGBlitavg	average value from literature for P in BGB	2.78	g P/m^2	Boar, 2006
POC_hydrolysis_constant	hydrolysis rate for particulate organic carbon	1*10^{-4}	day^{-1}	estimate
POC_settling_rate	settling rate for particulate organic carbon	0.05	day-1	van Dam *et al.,* 2007
PON_conc_lake	PON concentration in lake	0.5	g N/m^3	estimate

PON_conc_river	PON concentration in inflow	1	g N/m^3	estimate
PON_hydrolysis_constant	hydrolysis rate for particulate organic nitrogen	0.00063	day^{-1}	estimate
PON_settling_constant	settling rate for particulate organic nitrogen	0.05	day-1	van Dam *et al.*, 2007
PONP_conc	concentration of particulate organic nitrogen in pore water	IF(pore_water>0)THEN (PONP/pore_water)ELSE(0)	g N/m^3	
PONS_conc	concentration of particulate organic nitrogen in surface water	IF(surface_water>0)THEN (PONS/surface_water)ELSE(0)	g N/m^3	
POP_conc_lake	POP concentration in lake	0.1	g P/m^3	estimate
POP_conc_river	POP concentration in inflow	0.1	g P/m^3	estimate
POP_hydrolysis_constant	hydrolysis rate for particulate organic phosphorus	0.00063	day^{-1}	estimate
POP_settling_constant	settling rate for particulate organic phosphorus	0.05	day-1	estimate
POPP_conc	concentration of particulate organic phosphorus in pore water	IF(pore_water>0)THEN (POPP/pore_water)ELSE(0)	g P/m^3	
POPS_conc	concentration of particulate organic phosphorus in surface water	IF(surface_water>0)THEN(POPS /surface_water)ELSE(0)	g P/m^3	
pore_water	pore water	Water-surface_water	m^3/m^2	
porewater_free	pore water minus pore water at field capacity	IF(water_fraction>porosity) THEN((1-water_filled_porosity_field_capacity)*porosity*soil_volume) ELSE IF(water_fraction>water_filled_porosity_field_capacity*porosity) AND(water_fraction<porosity) OR(water_fraction=porosity) THEN((water_fraction-water_filled_porosity_field_capacity*porosity)*soil_volume) ELSE(0)	m^3/m^2	
porosity	porosity of the soil	0.8	-	estimate
radiance	mean monthly values for Naivasha	Counter (0, 365)	MJ/m^2*day	Muthuri *et al.*, 1989
rainfall_rate	mean monthly values for Naivasha	Counter (0, 365)	mm/day	Gaudet, 1978
river_inflow_rate	monthly means	Counter (0, 365)	m^3/day	calibrated
soil_depth	soil depth, rooting depth	0.2	m	estimate
soil_volume	soil volume	soil_depth*1	m^3/m^2	
surface_water	amount of surface water	IF(water_fraction<porosity) OR(water_fraction=porosity) THEN(0) ELSE((water_fraction-porosity)*soil_volume)	m^3/m^2	
tldownmax_C	max translocation rate for carbon from AGB to BGB	15	g C/day	calibrated
tldownmax_N	max translocation rate for nitrogen from AGB to BGB	0.5	g N/day	calibrated

tldownmax_P	max translocation rate for phosphorus from AGB to BGB	0.5		g P/day	calibrated
tlupmax_C	max translocation rate for carbon from BGB to AGB	0.5		g C/day	calibrated
tlupmax_N	max translocation rate for nitrogen from BGB to AGB	5		g N/day	calibrated
tlupmax_P	max translocation rate for phosphorus from BGB to AGB	1		g P/day	calibrated
water_filled_porosity	water filled porosity	water_fraction/porosity		-	
water_filled_porosity_field_capacity	water filled porosity at field capacity	0.625		-	estimate
water_fraction	volume fraction of pore water	Water/soil_volume		-	
wet_yes_or_no	switch to allow lake inflow or not	0 or 1		-	

Appendix 5.4: Transcript of equation layer Stella

Top-Level Model:
CAGB(t) = CAGB(t - dt) + (CAGB_assimilation - CAGB_death - C_translocation - CAGB_respiration - CAGB_harvesting) * dt
  INIT CAGB = 1853
  INFLOWS:
    CAGB_assimilation = max_assimilation_constant*CAGB*(1-(CAGB/(C_conc_AGB_lit*K_assim)))*limit_NP_ass*limit_radiance
  OUTFLOWS:
    CAGB_death = CAGB*CAGB_death_constant
    C_translocation = IF(CAGB>0)AND(CBGB>0) AND (C_AGB_to_BGB_optimal_ratio<(CAGB/CBGB)) THEN (tldownmax_C*((CAGB/CBGB)/((CAGB/CBGB)+Ktldown_C))) ELSE IF (CAGB>=0)AND (CBGB>0) AND (C_AGB_to_BGB_optimal_ratio>(CAGB/CBGB)) THEN (-tlupmax_C*((1)/((1)+Ktlup_C*CAGB/CBGB))) ELSE(0)
    CAGB_respiration = CAGB*maint_coeff_AGB+CAGB_assimilation*growth_coeff_AGB
    CAGB_harvesting = IF (harvest_batch_yes=1) AND (harvest_regular_yes=0) THEN (PULSE(harvest_%_of_AGB*CAGB, harvest_day_batch,harvest_interval_batch)) ELSE IF (harvest_regular_yes=1) AND (harvest_batch_yes=0) THEN (PULSE(harvest_in_g_CAGB, harvest_day_regular,harvest_interval_regular)) ELSE IF (harvest_batch_yes=1) AND (harvest_regular_yes=1) THEN (PULSE(harvest_%_of_AGB*CAGB, harvest_day_batch,harvest_interval_batch) + PULSE(harvest_in_g_CAGB, harvest_day_regular,harvest_interval_regular)) ELSE(0)
CBGB(t) = CBGB(t - dt) + (C_translocation - CBGB_death - CBGB_respiration) * dt
  INIT CBGB = 1570
  INFLOWS:
    C_translocation = IF(CAGB>0)AND(CBGB>0) AND (C_AGB_to_BGB_optimal_ratio<(CAGB/CBGB)) THEN (tldownmax_C*((CAGB/CBGB)/((CAGB/CBGB)+Ktldown_C))) ELSE IF (CAGB>=0)AND (CBGB>0) AND (C_AGB_to_BGB_optimal_ratio>(CAGB/CBGB)) THEN (-tlupmax_C*((1)/((1)+Ktlup_C*CAGB/CBGB))) ELSE(0)
  OUTFLOWS:
    CBGB_death = IF(CNBGB_ratio>75)THEN(CAGB_death_constant*CBGB)ELSE(CBGB*CBGB_death_constant)
    CBGB_respiration = IF(C_translocation>0)THEN(CBGB*maint_coeff_BGB+C_translocation*growth_coeff_BGB)ELSE(CBGB*maint_coeff_BGB)
CDAGB(t) = CDAGB(t - dt) + (CAGB_death - CDAGB_fragmentation - CDAGB_leaching) * dt
  INIT CDAGB = 335
  INFLOWS:
    CAGB_death = CAGB*CAGB_death_constant
  OUTFLOWS:
    CDAGB_fragmentation = IF(surface_water>=Ksurf)THEN(CDAGB*CDAGB_frag_constant)ELSE IF(surface_water<Ksurf)AND(surface_water>0) THEN((surface_water/Ksurf)*CDAGB*CDAGB_frag_constant)ELSE (0)
    CDAGB_leaching = IF(surface_water>=Ksurf)THEN(CDAGB_leach_constant*CAGB_death)ELSE IF(surface_water<Ksurf)AND(surface_water>0) THEN((surface_water/Ksurf)*CDAGB_leach_constant*CAGB_death)ELSE (0)
CDBGB(t) = CDBGB(t - dt) + (CBGB_death - CDBGB_fragmentation - CDBGB_leaching) * dt
  INIT CDBGB = 284
  INFLOWS:
    CBGB_death = IF(CNBGB_ratio>75)THEN(CAGB_death_constant*CBGB)ELSE(CBGB*CBGB_death_constant)
  OUTFLOWS:
    CDBGB_fragmentation = IF(pore_water>=Kpor)THEN(CDBGB*CDBGB_frag_constant)ELSE IF(pore_water<Kpor)AND(pore_water>0) THEN((pore_water/Kpor)*CDBGB*CDBGB_frag_constant)ELSE (0)
    CDBGB_leaching = IF(pore_water>=Kpor)THEN(CDBGB_leach_constant*CBGB_death)ELSE IF(pore_water<Kpor)AND(pore_water>0) THEN((pore_water/Kpor)*CDBGB_leach_constant*CBGB_death)ELSE (0)
DONP(t) = DONP(t - dt) + (PONP_hydrolysis + DON_diffusion + DONP_inflow_river + DONSP_drain - DONP_mineral - DONP_recharge) * dt
  INIT DONP = 1.1
  INFLOWS:

PONP_hydrolysis = IF(pore_water>=Kpor)THEN(PONP*PON_hydrolysis_const)ELSE
IF(pore_water<Kpor)AND(pore_water>0) THEN((pore_water/Kpor)*PONP*PON_hydrolysis_const)ELSE (0)
    DON_diffusion = IF(surface_water>0) THEN(K_DON_diffusion*((DONS/surface_water)-
(DONP/soil_depth))/((surface_water+soil_depth)/2)) ELSE(0)
    DONP_inflow_river = IF(surface_water=0)THEN(DON_conc_river*river_inflow)ELSE
IF(surface_water<Ksurf)AND(surface_water>0)THEN ((1-surface_water/Ksurf)*(DON_conc_river*river_inflow))
ELSE (0)
    DONSP_drain = IF(surface_water=0)THEN(DONS)ELSE IF(surface_water<Ksurf)AND(surface_water>0)
THEN(DONS*(1-surface_water/Ksurf)) ELSE(0)
  OUTFLOWS:
    DONP_mineral = IF(pore_water>=Kpor)THEN(DON_mineral_const*DONP)ELSE
IF(pore_water<Kpor)AND(pore_water>0) THEN((pore_water/Kpor)*DON_mineral_const*DONP)ELSE (0)
    DONP_recharge = DONP_conc*recharge
DONS(t) = DONS(t - dt) + (PONS_hydrolysis + DONS_inflow_river + DON_inflow_lake - DONS_mineral -
DON_diffusion - DONS_outflow - DONSP_drain) * dt
  INIT DONS = 1.1
  INFLOWS:
    PONS_hydrolysis = IF(surface_water>=Ksurf)THEN(PONS*PON_hydrolysis_const)ELSE
IF(surface_water<Ksurf)AND(surface_water>0) THEN((surface_water/Ksurf)*PONS*PON_hydrolysis_const)ELSE
(0)
    DONS_inflow_river = IF(surface_water>=Ksurf) THEN(DON_conc_river*river_inflow) ELSE
IF(surface_water<Ksurf)AND(surface_water>0)THEN (surface_water/Ksurf*DON_conc_river*river_inflow) ELSE
(0)
    DON_inflow_lake = inflow_lake*DON_conc_lake
  OUTFLOWS:
    DONS_mineral = IF(surface_water>=Ksurf)THEN(DON_mineral_const*DONS)ELSE
IF(surface_water<Ksurf)AND(surface_water>0) THEN((surface_water/Ksurf)*DON_mineral_const*DONS)ELSE (0)
    DON_diffusion = IF(surface_water>0) THEN(K_DON_diffusion*((DONS/surface_water)-
(DONP/soil_depth))/((surface_water+soil_depth)/2)) ELSE(0)
    DONS_outflow = outflow*DONS_conc
    DONSP_drain = IF(surface_water=0)THEN(DONS)ELSE IF(surface_water<Ksurf)AND(surface_water>0)
THEN(DONS*(1-surface_water/Ksurf)) ELSE(0)
DOPP(t) = DOPP(t - dt) + (DOP_diffusion + POPP_hydrolysis + DOPP_inflow_river + DOPSP_drain - DOPP_recharge
- DOPP_mineral) * dt
  INIT DOPP = 1
  INFLOWS:
    DOP_diffusion = IF(surface_water>0)THEN(K_DOP_diffusion*((DOPS/surface_water)-
(DOPP/soil_depth))/((surface_water+soil_depth)/2)) ELSE(0)
    POPP_hydrolysis = IF(pore_water>=Kpor)THEN(POPP*POP_hydrolysis_constant)ELSE
IF(pore_water<Kpor)AND(pore_water>0) THEN((pore_water/Kpor)*POPP*POP_hydrolysis_constant)ELSE (0)
    DOPP_inflow_river = IF(surface_water=0)THEN(DOP_conc_river*river_inflow)ELSE
IF(surface_water<Ksurf)AND(surface_water>0)THEN ((1-surface_water/Ksurf)*(DOP_conc_river*river_inflow))
ELSE (0)
    DOPSP_drain = IF(surface_water=0)THEN(DOPS)ELSE IF(surface_water<Ksurf)AND(surface_water>0)
THEN(DOPS*(1-surface_water/Ksurf)) ELSE(0)
  OUTFLOWS:
    DOPP_recharge = DOPP_conc*recharge
    DOPP_mineral = IF(pore_water>=Kpor)THEN(DOP_mineral_const*DOPP)ELSE
IF(pore_water<Kpor)AND(pore_water>0) THEN((pore_water/Kpor)*DOP_mineral_const*DOPP)ELSE (0)
DOPS(t) = DOPS(t - dt) + (POPS_hydrolysis + DOP_inflow_lake + DOPS_inflow_river - DOPS_mineral -
DOPS_outflow - DOP_diffusion - DOPSP_drain) * dt
  INIT DOPS = 1
  INFLOWS:
    POPS_hydrolysis = IF(surface_water>=Ksurf)THEN(POPS*POP_hydrolysis_constant)ELSE
IF(surface_water<Ksurf)AND(surface_water>0)
THEN((surface_water/Ksurf)*POPS*POP_hydrolysis_constant)ELSE (0)
    DOP_inflow_lake = DOP_conc_lake*inflow_lake

DOPS_inflow_river = IF(surface_water>=Ksurf) THEN(DOP_conc_river*river_inflow) ELSE IF(surface_water<Ksurf)AND(surface_water>0)THEN (surface_water/Ksurf*DOP_conc_river*river_inflow) ELSE (0)

  OUTFLOWS:

    DOPS_mineral = IF(surface_water>=Ksurf)THEN(DOP_mineral_const*DOPS)ELSE IF(surface_water<Ksurf)AND(surface_water>0) THEN((surface_water/Ksurf)*DOP_mineral_const*DOPS)ELSE (0)

    DOPS_outflow = DOPS_conc*outflow

    DOP_diffusion = IF(surface_water>0)THEN(K_DOP_diffusion*((DOPS/surface_water)-(DOPP/soil_depth))/((surface_water+soil_depth)/2)) ELSE(0)

    DOPSP_drain = IF(surface_water=0)THEN(DOPS)ELSE IF(surface_water<Ksurf)AND(surface_water>0) THEN(DOPS*(1-surface_water/Ksurf)) ELSE(0)

NAGB(t) = NAGB(t - dt) + (- N_translocation - N_retrans - NAGB_death - NAGB_harvesting) * dt

  INIT NAGB = 44

  OUTFLOWS:

    N_translocation = IF(NAGB>=0)AND(NBGB>0) AND (N_AGB_to_BGB_optimal_ratio<(NAGB/NBGB))AND (N_conc_BGB<=N_conc_BGB_lit) THEN (tldownmax_N*((NAGB/NBGB)/((NAGB/NBGB)+Ktldown_N))) ELSE IF (NAGB>=0)AND (NBGB>0) AND (N_AGB_to_BGB_optimal_ratio>(NAGB/NBGB)) AND (N_conc_AGB<=N_conc_AGB_lit) THEN (-tlupmax_N*(1/(1+Ktlup_N*(NAGB/NBGB)))) ELSE(0)

    N_retrans = IF(CNAGB_ratio>0)THEN(CAGB_death*N_retrans_constant/CNAGB_ratio)ELSE(0)

    NAGB_death = IF(CNAGB_ratio>0)THEN(CAGB_death*(1-N_retrans_constant)/CNAGB_ratio)ELSE(0)

    NAGB_harvesting = IF(CNAGB_ratio>0)THEN(CAGB_harvesting/CNAGB_ratio)ELSE(0)

NBGB(t) = NBGB(t - dt) + (N_translocation + N_retrans + NH4_uptake + NO3_uptake - NBGB_death) * dt

  INIT NBGB = 31

  INFLOWS:

    N_translocation = IF(NAGB>=0)AND(NBGB>0) AND (N_AGB_to_BGB_optimal_ratio<(NAGB/NBGB))AND (N_conc_BGB<=N_conc_BGB_lit) THEN (tldownmax_N*((NAGB/NBGB)/((NAGB/NBGB)+Ktldown_N))) ELSE IF (NAGB>=0)AND (NBGB>0) AND (N_AGB_to_BGB_optimal_ratio>(NAGB/NBGB)) AND (N_conc_AGB<=N_conc_AGB_lit) THEN (-tlupmax_N*(1/(1+Ktlup_N*(NAGB/NBGB)))) ELSE(0)

    N_retrans = IF(CNAGB_ratio>0)THEN(CAGB_death*N_retrans_constant/CNAGB_ratio)ELSE(0)

    NH4_uptake = (IF(NBGB/BGB<=N_conc_BGB_lit)THEN(max_NH4_uptake*(NAGB+NBGB)*(1-(NAGB+NBGB)/(N_max_AGB+N_max_BGB))*limit_NH4_uptake)ELSE(0))

    NO3_uptake = IF(NBGB/BGB<=N_conc_BGB_lit)THEN(max_NO3_uptake*(NAGB+NBGB)*(1-(NAGB+NBGB)/(N_max_AGB+N_max_BGB))*limit_NO3_uptake)ELSE(0)

  OUTFLOWS:

    NBGB_death = IF(CNBGB_ratio>0) THEN(CBGB_death/CNBGB_ratio) ELSE (0)

NDAGB(t) = NDAGB(t - dt) + (NAGB_death - NDAGB_fragmentation - NDAGB_leaching) * dt

  INIT NDAGB = 7.9

  INFLOWS:

    NAGB_death = IF(CNAGB_ratio>0)THEN(CAGB_death*(1-N_retrans_constant)/CNAGB_ratio)ELSE(0)

  OUTFLOWS:

    NDAGB_fragmentation = IF(surface_water>=Ksurf)THEN(CDAGB_fragmentation/CNDAGB_ratio)ELSE IF(surface_water<Ksurf)AND(surface_water>0) THEN((surface_water/Ksurf)*CDAGB_fragmentation/CNDAGB_ratio)ELSE (0)

    NDAGB_leaching = IF(surface_water>=Ksurf)THEN(CDAGB_leaching/CNDAGB_ratio)ELSE IF(surface_water<Ksurf)AND(surface_water>0) THEN((surface_water/Ksurf)*CDAGB_leaching/CNDAGB_ratio)ELSE (0)

NDBGB(t) = NDBGB(t - dt) + (NBGB_death - NDBGB_fragmentation - NDBGB_leaching) * dt

  INIT NDBGB = 5.6

  INFLOWS:

    NBGB_death = IF(CNBGB_ratio>0) THEN(CBGB_death/CNBGB_ratio) ELSE (0)

  OUTFLOWS:

    NDBGB_fragmentation = IF(pore_water>=Kpor)THEN(CDBGB_fragmentation/CNDBGB_ratio)ELSE IF(pore_water<Kpor)AND(pore_water>0) THEN((pore_water/Kpor)*CDBGB_fragmentation/CNDBGB_ratio)ELSE (0)

    NDBGB_leaching = IF(pore_water>=Kpor)THEN(CDBGB_leaching/CNDBGB_ratio)ELSE IF(pore_water<Kpor)AND(pore_water>0) THEN((pore_water/Kpor)*CDBGB_leaching/CNDBGB_ratio)ELSE (0)

135

NH4P(t) = NH4P(t - dt) + (DONP_mineral + NH4_diffusion + NDBGB_leaching + NH4P_inflow_river + NH4SP_drain - nitrification_P - NH4_uptake - NH4P_recharge) * dt

   INIT NH4P = 0.5

   INFLOWS:

     DONP_mineral = IF(pore_water>=Kpor)THEN(DON_mineral_const*DONP)ELSE IF(pore_water<Kpor)AND(pore_water>0) THEN((pore_water/Kpor)*DON_mineral_const*DONP)ELSE (0)

     NH4_diffusion = IF(surface_water>0) THEN(K_NH4_diffusion*((NH4S/surface_water)-(NH4P/soil_depth))/((surface_water+soil_depth)/2)) ELSE(0)

     NDBGB_leaching = IF(pore_water>=Kpor)THEN(CDBGB_leaching/CNDBGB_ratio)ELSE IF(pore_water<Kpor)AND(pore_water>0) THEN((pore_water/Kpor)*CDBGB_leaching/CNDBGB_ratio)ELSE (0)

     NH4P_inflow_river = IF(surface_water=0)THEN(NH4_conc_river*river_inflow)ELSE IF(surface_water<Ksurf)AND(surface_water>0)THEN ((1-surface_water/Ksurf)*(NH4_conc_river*river_inflow)) ELSE (0)

     NH4SP_drain = IF(surface_water=0)THEN(NH4S)ELSE IF(surface_water<Ksurf)AND(surface_water>0) THEN(NH4S*(1-surface_water/Ksurf)) ELSE(0)

   OUTFLOWS:

     nitrification_P = IF(pore_water>=Kpor)THEN(K_nitri*mode*NH4P)ELSE IF(pore_water<Kpor)AND(pore_water>0) THEN((pore_water/Kpor)*(K_nitri*mode*NH4P))ELSE (0)

     NH4_uptake = (IF(NBGB/BGB<=N_conc_BGB_lit)THEN(max_NH4_uptake*(NAGB+NBGB)*(1-(NAGB+NBGB)/(N_max_AGB+N_max_BGB))*limit_NH4_uptake)ELSE(0))

     NH4P_recharge = NH4P_conc*recharge

NH4S(t) = NH4S(t - dt) + (DONS_mineral + NH4S_inflow_river + NH4_inflow_lake + NDAGB_leaching - nitrification_S - NH4_diffusion - NH4S_outflow - NH4SP_drain) * dt

   INIT NH4S = 0.5

   INFLOWS:

     DONS_mineral = IF(surface_water>=Ksurf)THEN(DON_mineral_const*DONS)ELSE IF(surface_water<Ksurf)AND(surface_water>0) THEN((surface_water/Ksurf)*DON_mineral_const*DONS)ELSE (0)

     NH4S_inflow_river = IF(surface_water>=Ksurf) THEN(NH4_conc_river*river_inflow) ELSE IF(surface_water<Ksurf)AND(surface_water>0)THEN (surface_water/Ksurf*NH4_conc_river*river_inflow) ELSE (0)

     NH4_inflow_lake = inflow_lake*NH4_conc_lake

     NDAGB_leaching = IF(surface_water>=Ksurf)THEN(CDAGB_leaching/CNDAGB_ratio)ELSE IF(surface_water<Ksurf)AND(surface_water>0) THEN((surface_water/Ksurf)*CDAGB_leaching/CNDAGB_ratio)ELSE (0)

   OUTFLOWS:

     nitrification_S = IF(surface_water>=Ksurf)THEN(K_nitri*NH4S)ELSE IF(surface_water<Ksurf)AND(surface_water>0) THEN((surface_water/Ksurf)*(K_nitri*NH4S))ELSE (0)

     NH4_diffusion = IF(surface_water>0) THEN(K_NH4_diffusion*((NH4S/surface_water)-(NH4P/soil_depth))/((surface_water+soil_depth)/2)) ELSE(0)

     NH4S_outflow = outflow*NH4S_conc

     NH4SP_drain = IF(surface_water=0)THEN(NH4S)ELSE IF(surface_water<Ksurf)AND(surface_water>0) THEN(NH4S*(1-surface_water/Ksurf)) ELSE(0)

NO3P(t) = NO3P(t - dt) + (nitrification_P + NO3_diffusion + NO3P_inflow_river + NO3SP_drain - NO3_uptake - NO3P_recharge - denitrification_P) * dt

   INIT NO3P = 0.05

   INFLOWS:

     nitrification_P = IF(pore_water>=Kpor)THEN(K_nitri*mode*NH4P)ELSE IF(pore_water<Kpor)AND(pore_water>0) THEN((pore_water/Kpor)*(K_nitri*mode*NH4P))ELSE (0)

     NO3_diffusion = IF(surface_water>0) THEN(K_NO3_diffusion*((NO3S/surface_water)-(NO3P/soil_depth))/((surface_water+soil_depth)/2)) ELSE(0)

     NO3P_inflow_river = IF(surface_water=0)THEN(NO3_conc_river*river_inflow)ELSE IF(surface_water<Ksurf)AND(surface_water>0)THEN ((1-surface_water/Ksurf)*(NO3_conc_river*river_inflow)) ELSE (0)

     NO3SP_drain = IF(surface_water=0)THEN(NO3S)ELSE IF(surface_water<Ksurf)AND(surface_water>0) THEN(NO3S*(1-surface_water/Ksurf)) ELSE(0)

   OUTFLOWS:

NO3_uptake = IF(NBGB/BGB<=N_conc_BGB_lit)THEN(max_NO3_uptake*(NAGB+NBGB)*(1-(NAGB+NBGB)/(N_max_AGB+N_max_BGB))*limit_NO3_uptake)ELSE(0)

NO3P_recharge = NO3P_conc*recharge

denitrification_P = (IF(pore_water>=Kpor)THEN(K_denitri*NO3P*(1-mode))ELSE IF(pore_water<Kpor)AND(pore_water>0) THEN((pore_water/Kpor)*(K_denitri*NO3P*(1-mode)))ELSE (0))

NO3S(t) = NO3S(t - dt) + (nitrification_S + NO3_inflow_river + NO3_inflow_lake - NO3_diffusion - NO3S_outflow - NO3SP_drain) * dt

  INIT NO3S = 0.05

  INFLOWS:

nitrification_S = IF(surface_water>=Ksurf)THEN(K_nitri*NH4S)ELSE IF(surface_water<Ksurf)AND(surface_water>0) THEN((surface_water/Ksurf)*(K_nitri*NH4S))ELSE (0)

NO3_inflow_river = IF(surface_water>=Ksurf) THEN(NO3_conc_river*river_inflow) ELSE IF(surface_water<Ksurf)AND(surface_water>0)THEN (surface_water/Ksurf*NO3_conc_river*river_inflow) ELSE (0)

NO3_inflow_lake = inflow_lake*NO3_conc_lake

  OUTFLOWS:

NO3_diffusion = IF(surface_water>0) THEN(K_NO3_diffusion*((NO3S/surface_water)-(NO3P/soil_depth))/((surface_water+soil_depth)/2)) ELSE(0)

NO3S_outflow = outflow*NO3S_conc

NO3SP_drain = IF(surface_water=0)THEN(NO3S)ELSE IF(surface_water<Ksurf)AND(surface_water>0) THEN(NO3S*(1-surface_water/Ksurf)) ELSE(0)

OPADS(t) = OPADS(t - dt) + (OP_adsorption - OP_desorption) * dt

  INIT OPADS = 1

  INFLOWS:

OP_adsorption = IF(OPADS<OPADS_eq)THEN((1-OPADS/OPADS_max)*(OPADS_eq-OPADS)) ELSE(0)

  OUTFLOWS:

OP_desorption = (IF(OPADS>OPADS_eq) THEN(OPADS/OPADS_max*(OPADS-OPADS_eq)) ELSE(0))

OPP(t) = OPP(t - dt) + (OP_diffusion + DOPP_mineral + OP_desorption + PDBGB_leaching + OPSP_drain + OPP_river_inflow - OP_uptake - OPP_recharge - OP_adsorption) * dt

  INIT OPP = 1

  INFLOWS:

OP_diffusion = IF(surface_water>0)THEN(K_OP_diffusion*((OPS/surface_water)-(OPP/soil_depth))/((soil_depth+surface_water)/2)) ELSE(0)

DOPP_mineral = IF(pore_water>=Kpor)THEN(DOP_mineral_const*DOPP)ELSE IF(pore_water<Kpor)AND(pore_water>0) THEN((pore_water/Kpor)*DOP_mineral_const*DOPP)ELSE (0)

OP_desorption = (IF(OPADS>OPADS_eq) THEN(OPADS/OPADS_max*(OPADS-OPADS_eq)) ELSE(0))

PDBGB_leaching = IF(pore_water>=Kpor)THEN(CDBGB_leaching/CPDBGB_ratio)ELSE IF(pore_water<Kpor)AND(pore_water>0) THEN((pore_water/Kpor)*CDBGB_leaching/CPDBGB_ratio)ELSE (0)

OPSP_drain = IF(surface_water=0)THEN(OPS)ELSE IF(surface_water<Ksurf)AND(surface_water>0) THEN(OPS*(1-surface_water/Ksurf)) ELSE(0)

OPP_river_inflow = IF(surface_water=0)THEN(OP_conc_river*river_inflow)ELSE IF(surface_water<Ksurf)AND(surface_water>0)THEN ((1-surface_water/Ksurf)*(OP_conc_river*river_inflow)) ELSE (0)

  OUTFLOWS:

OP_uptake = IF(PBGB/BGB<=P_conc_BGB_lit)THEN(max_OP_uptake*(PAGB+PBGB)*(1-(PAGB+PBGB)/(P_max_AGB+P_max_BGB))*limit_OP_uptake)ELSE(0)

OPP_recharge = OPP_conc*recharge

OP_adsorption = IF(OPADS<OPADS_eq)THEN((1-OPADS/OPADS_max)*(OPADS_eq-OPADS)) ELSE(0)

OPS(t) = OPS(t - dt) + (DOPS_mineral + PDAGB_leaching + OP_inflow_lake + OPS_inflow_river - OP_diffusion - OPS_outflow - OPSP_drain) * dt

  INIT OPS = 1

  INFLOWS:

DOPS_mineral = IF(surface_water>=Ksurf)THEN(DOP_mineral_const*DOPS)ELSE IF(surface_water<Ksurf)AND(surface_water>0) THEN((surface_water/Ksurf)*DOP_mineral_const*DOPS)ELSE (0)

PDAGB_leaching = IF(surface_water>=Ksurf)THEN(CDAGB_leaching/CPDAGB_ratio)ELSE IF(surface_water<Ksurf)AND(surface_water>0) THEN((surface_water/Ksurf)*CDAGB_leaching/CPDAGB_ratio)ELSE (0)

OP_inflow_lake = OP_conc_lake*inflow_lake

OPS_inflow_river = IF(surface_water>=Ksurf) THEN(OP_conc_river*river_inflow) ELSE
IF(surface_water<Ksurf)AND(surface_water>0)THEN (surface_water/Ksurf*OP_conc_river*river_inflow) ELSE (0)

  OUTFLOWS:

OP_diffusion = IF(surface_water>0)THEN(K_OP_diffusion*((OPS/surface_water)-
(OPP/soil_depth))/((soil_depth+surface_water)/2)) ELSE(0)

OPS_outflow = OPS_conc*outflow

OPSP_drain = IF(surface_water=0)THEN(OPS)ELSE IF(surface_water<Ksurf)AND(surface_water>0)
THEN(OPS*(1-surface_water/Ksurf)) ELSE(0)

PAGB(t) = PAGB(t - dt) + (- P_retrans - PAGB_death - P_translocation - PAGB_harvesting) * dt

  INIT PAGB = 2.61

  OUTFLOWS:

P_retrans = IF(CPAGB_ratio>0)THEN(P_retrans_constant*CAGB_death/CPAGB_ratio)ELSE(0)

PAGB_death = IF(CPAGB_ratio>0)THEN(CAGB_death*(1-P_retrans_constant)/CPAGB_ratio)ELSE(0)

P_translocation = IF(PAGB>=0)AND(PBGB>0) AND (P_AGB_to_BGB_optimal_ratio<(PAGB/PBGB))AND
(P_conc_BGB<=P_conc_BGB_lit) THEN (tldownmax_P*((PAGB/PBGB)/((PAGB/PBGB)+Ktldown_P))) ELSE IF
(PAGB>=0)AND (PBGB>0) AND (P_AGB_to_BGB_optimal_ratio>(PAGB/PBGB)) AND
(P_conc_AGB<=P_conc_AGB_lit) THEN (-tlupmax_P*(1/(1+Ktlup_P*(PAGB/PBGB)))) ELSE(0)

PAGB_harvesting = IF(CPAGB_ratio>0)THEN(CAGB_harvesting/CPAGB_ratio)ELSE(0)

PBGB(t) = PBGB(t - dt) + (P_retrans + OP_uptake + P_translocation - PBGB_death) * dt

  INIT PBGB = 2.78

  INFLOWS:

P_retrans = IF(CPAGB_ratio>0)THEN(P_retrans_constant*CAGB_death/CPAGB_ratio)ELSE(0)

OP_uptake = IF(PBGB/BGB<=P_conc_BGB_lit)THEN(max_OP_uptake*(PAGB+PBGB)*(1-
(PAGB+PBGB)/(P_max_AGB+P_max_BGB))*limit_OP_uptake)ELSE(0)

P_translocation = IF(PAGB>=0)AND(PBGB>0) AND (P_AGB_to_BGB_optimal_ratio<(PAGB/PBGB))AND
(P_conc_BGB<=P_conc_BGB_lit) THEN (tldownmax_P*((PAGB/PBGB)/((PAGB/PBGB)+Ktldown_P))) ELSE IF
(PAGB>=0)AND (PBGB>0) AND (P_AGB_to_BGB_optimal_ratio>(PAGB/PBGB)) AND
(P_conc_AGB<=P_conc_AGB_lit) THEN (-tlupmax_P*(1/(1+Ktlup_P*(PAGB/PBGB)))) ELSE(0)

  OUTFLOWS:

PBGB_death = IF(CPBGB_ratio>0) THEN(CBGB_death/CPBGB_ratio) ELSE (0)

PDAGB(t) = PDAGB(t - dt) + (PAGB_death - PDAGB_fragmentation - PDAGB_leaching) * dt

  INIT PDAGB = 0.47

  INFLOWS:

PAGB_death = IF(CPAGB_ratio>0)THEN(CAGB_death*(1-P_retrans_constant)/CPAGB_ratio)ELSE(0)

  OUTFLOWS:

PDAGB_fragmentation = IF(surface_water>=Ksurf)THEN(CDAGB_fragmentation/CPDAGB_ratio)ELSE
IF(surface_water<Ksurf)AND(surface_water>0)
THEN((surface_water/Ksurf)*CDAGB_fragmentation/CPDAGB_ratio)ELSE (0)

PDAGB_leaching = IF(surface_water>=Ksurf)THEN(CDAGB_leaching/CPDAGB_ratio)ELSE
IF(surface_water<Ksurf)AND(surface_water>0)
THEN((surface_water/Ksurf)*CDAGB_leaching/CPDAGB_ratio)ELSE (0)

PDBGB(t) = PDBGB(t - dt) + (PBGB_death - PDBGB_fragmentation - PDBGB_leaching) * dt

  INIT PDBGB = 0.5

  INFLOWS:

PBGB_death = IF(CPBGB_ratio>0) THEN(CBGB_death/CPBGB_ratio) ELSE (0)

  OUTFLOWS:

PDBGB_fragmentation = IF(pore_water>=Kpor)THEN(CDBGB_fragmentation/CPDBGB_ratio)ELSE
IF(pore_water<Kpor)AND(pore_water>0) THEN((pore_water/Kpor)*CDBGB_fragmentation/CPDBGB_ratio)ELSE
(0)

PDBGB_leaching = IF(pore_water>=Kpor)THEN(CDBGB_leaching/CPDBGB_ratio)ELSE
IF(pore_water<Kpor)AND(pore_water>0) THEN((pore_water/Kpor)*CDBGB_leaching/CPDBGB_ratio)ELSE (0)

POCP(t) = POCP(t - dt) + (CDBGB_fragmentation + POC_settling + POC_drain - POCP_hydrolysis) * dt

  INIT POCP = 20

  INFLOWS:

CDBGB_fragmentation = IF(pore_water>=Kpor)THEN(CDBGB*CDBGB_frag_constant)ELSE
IF(pore_water<Kpor)AND(pore_water>0) THEN((pore_water/Kpor)*CDBGB*CDBGB_frag_constant)ELSE (0)

POC_settling = POC_settling_rate*POCS

POC_drain = IF(surface_water=0)THEN(POCS)ELSE IF(surface_water<Ksurf)AND(surface_water>0) THEN(POCS*(1-surface_water/Ksurf)) ELSE(0)

OUTFLOWS:

POCP_hydrolysis = IF(pore_water>=Kpor)THEN(POCP*POC_hydrolysis_constant)ELSE IF(pore_water<Kpor)AND(pore_water>0) THEN((pore_water/Kpor)*POCP*POC_hydrolysis_constant)ELSE (0)

POCS(t) = POCS(t - dt) + (CDAGB_fragmentation - POCS_hydrolysis - POC_settling - POC_drain) * dt

INIT POCS = 20

INFLOWS:

CDAGB_fragmentation = IF(surface_water>=Ksurf)THEN(CDAGB*CDAGB_frag_constant)ELSE IF(surface_water<Ksurf)AND(surface_water>0) THEN((surface_water/Ksurf)*CDAGB*CDAGB_frag_constant)ELSE (0)

OUTFLOWS:

POCS_hydrolysis = IF(surface_water>=Ksurf)THEN(POCS*POC_hydrolysis_constant)ELSE IF(surface_water<Ksurf)AND(surface_water>0) THEN((surface_water/Ksurf)*POCS*POC_hydrolysis_constant)ELSE (0)

POC_settling = POC_settling_rate*POCS

POC_drain = IF(surface_water=0)THEN(POCS)ELSE IF(surface_water<Ksurf)AND(surface_water>0) THEN(POCS*(1-surface_water/Ksurf)) ELSE(0)

PONP(t) = PONP(t - dt) + (NDBGB_fragmentation + PON_settling + PONSP_drain + PONP_inflow_river - PONP_hydrolysis - PONP_recharge) * dt

INIT PONP = 0.9

INFLOWS:

NDBGB_fragmentation = IF(pore_water>=Kpor)THEN(CDBGB_fragmentation/CNDBGB_ratio)ELSE IF(pore_water<Kpor)AND(pore_water>0) THEN((pore_water/Kpor)*CDBGB_fragmentation/CNDBGB_ratio)ELSE (0)

PON_settling = PONS*PON_settling_constant

PONSP_drain = IF(surface_water=0)THEN(PONS)ELSE IF(surface_water<Ksurf)AND(surface_water>0) THEN(PONS*(1-surface_water/Ksurf)) ELSE(0)

PONP_inflow_river = IF(surface_water=0)THEN(PON_conc_river*river_inflow)ELSE IF(surface_water<Ksurf)AND(surface_water>0)THEN ((1-surface_water/Ksurf)*(PON_conc_river*river_inflow)) ELSE (0)

OUTFLOWS:

PONP_hydrolysis = IF(pore_water>=Kpor)THEN(PONP*PON_hydrolysis_const)ELSE IF(pore_water<Kpor)AND(pore_water>0) THEN((pore_water/Kpor)*PONP*PON_hydrolysis_const)ELSE (0)

PONP_recharge = PONP_conc*recharge

PONS(t) = PONS(t - dt) + (NDAGB_fragmentation + PONS_inflow_river + PON_inflow_lake - PONS_hydrolysis - PON_settling - PONS_outflow - PONSP_drain) * dt

INIT PONS = 0.9

INFLOWS:

NDAGB_fragmentation = IF(surface_water>=Ksurf)THEN(CDAGB_fragmentation/CNDAGB_ratio)ELSE IF(surface_water<Ksurf)AND(surface_water>0) THEN((surface_water/Ksurf)*CDAGB_fragmentation/CNDAGB_ratio)ELSE (0)

PONS_inflow_river = IF(surface_water>=Ksurf) THEN(PON_conc_river*river_inflow) ELSE IF(surface_water<Ksurf)AND(surface_water>0)THEN (surface_water/Ksurf*PON_conc_river*river_inflow) ELSE (0)

PON_inflow_lake = inflow_lake*PON_conc_lake

OUTFLOWS:

PONS_hydrolysis = IF(surface_water>=Ksurf)THEN(PONS*PON_hydrolysis_const)ELSE IF(surface_water<Ksurf)AND(surface_water>0) THEN((surface_water/Ksurf)*PONS*PON_hydrolysis_const)ELSE (0)

PON_settling = PONS*PON_settling_constant

PONS_outflow = outflow*PONS_conc

PONSP_drain = IF(surface_water=0)THEN(PONS)ELSE IF(surface_water<Ksurf)AND(surface_water>0) THEN(PONS*(1-surface_water/Ksurf)) ELSE(0)

POPP(t) = POPP(t - dt) + (POP_settling + PDBGB_fragmentation + POPP_inflow_river + POPSP_drain - POPP_recharge - POPP_hydrolysis) * dt

INIT POPP = 1

INFLOWS:

POP_settling = POPS*POP_settling_constant

PDBGB_fragmentation = IF(pore_water>=Kpor)THEN(CDBGB_fragmentation/CPDBGB_ratio)ELSE IF(pore_water<Kpor)AND(pore_water>0) THEN((pore_water/Kpor)*CDBGB_fragmentation/CPDBGB_ratio)ELSE (0)

POPP_inflow_river = IF(surface_water=0)THEN(POP_conc_river*river_inflow)ELSE IF(surface_water<Ksurf)AND(surface_water>0)THEN ((1-surface_water/Ksurf)*(POP_conc_river*river_inflow)) ELSE (0)

POPSP_drain = IF(surface_water=0)THEN(POPS)ELSE IF(surface_water<Ksurf)AND(surface_water>0) THEN(POPS*(1-surface_water/Ksurf)) ELSE(0)

OUTFLOWS:

POPP_recharge = POPP_conc*recharge

POPP_hydrolysis = IF(pore_water>=Kpor)THEN(POPP*POP_hydrolysis_constant)ELSE IF(pore_water<Kpor)AND(pore_water>0) THEN((pore_water/Kpor)*POPP*POP_hydrolysis_constant)ELSE (0)

POPS(t) = POPS(t - dt) + (PDAGB_fragmentation + POP_inflow_lake + POPS_inflow_river - POPS_hydrolysis - POPS_outflow - POP_settling - POPSP_drain) * dt

INIT POPS = 1

INFLOWS:

PDAGB_fragmentation = IF(surface_water>=Ksurf)THEN(CDAGB_fragmentation/CPDAGB_ratio)ELSE IF(surface_water<Ksurf)AND(surface_water>0) THEN((surface_water/Ksurf)*CDAGB_fragmentation/CPDAGB_ratio)ELSE (0)

POP_inflow_lake = inflow_lake*POP_conc_lake

POPS_inflow_river = IF(surface_water>=Ksurf) THEN(POP_conc_river*river_inflow) ELSE IF(surface_water<Ksurf)AND(surface_water>0)THEN (surface_water/Ksurf*POP_conc_river*river_inflow) ELSE (0)

OUTFLOWS:

POPS_hydrolysis = IF(surface_water>=Ksurf)THEN(POPS*POP_hydrolysis_constant)ELSE IF(surface_water<Ksurf)AND(surface_water>0) THEN((surface_water/Ksurf)*POPS*POP_hydrolysis_constant)ELSE (0)

POPS_outflow = outflow*POPS_conc

POP_settling = POPS*POP_settling_constant

POPSP_drain = IF(surface_water=0)THEN(POPS)ELSE IF(surface_water<Ksurf)AND(surface_water>0) THEN(POPS*(1-surface_water/Ksurf)) ELSE(0)

Water(t) = Water(t - dt) + (precipitation + inflow_lake + river_inflow - recharge - outflow - evaporation) * dt

INIT Water = 0.2

INFLOWS:

precipitation = rainfall_rate*0.001

inflow_lake = (IF(surface_water<max_depth) THEN(lake_inflow_rate) ELSE(0))*wet_yes_or_no

river_inflow = river_inflow_rate

OUTFLOWS:

recharge = fraction_out*porewater_free

outflow = IF(surface_water>max_depth) THEN(river_inflow+precipitation-evaporation-recharge) ELSE(0)

evaporation = evaporation_rate*0.001

AGB = CAGB/C_conc_AGB_lit

AGB_total_biomass_ratio = AGBlitavg/papyrus_biomass_litavg

AGBlitavg = 3489

BGB = CBGB/C_conc_BGB_lit

BGB_total_biomass_ratio = BGBlitavg/papyrus_biomass_litavg

BGBlitavg = 3928

bulkdens = 180000

C_AGB_to_BGB_optimal_ratio = CAGBlitavg/CBGBlitavg

C_conc_AGB = IF(AGB>0)THEN(CAGB/AGB)ELSE(0)

C_conc_AGB_lit = CAGBlitavg/AGBlitavg

C_conc_BGB = IF (BGB>0) THEN(CBGB/BGB) ELSE(0)

C_conc_BGB_lit = CBGBlitavg/BGBlitavg

C_conc_papyrus = IF(papyrus_biomass>0)THEN((CAGB+CBGB)/papyrus_biomass)ELSE(0)

C_initial_DAGB = papyrus_initial_dead_biomass*AGB_total_biomass_ratio*C_conc_AGB_lit

C_initial_DBGB = papyrus_initial_dead_biomass*BGB_total_biomass_ratio*C_conc_BGB_lit

C_max_AGB = papyrus_max_biomass*C_conc_AGB_lit*AGB_total_biomass_ratio

C_max_BGB = papyrus_max_biomass*C_conc_BGB_lit*BGB_total_biomass_ratio

C_papyrus = CAGB+CBGB

CAGB_death_constant = 0.0057

CAGBlitavg = 1853

CBGB_death_constant = 0.0014

CBGBlitavg = 1570

CDAGB_frag_constant = 0.0005

CDAGB_leach_constant = 0.432

CDBGB_frag_constant = 0.000834285714285714

CDBGB_leach_constant = 0.486

CNAGB_ratio = IF (NAGB>0.00001) THEN(CAGB/NAGB) ELSE(0)

CNBGB_ratio = IF(NBGB>0.00001)THEN(CBGB/NBGB)ELSE(0)

CNDAGB_ratio = CDAGB/NDAGB

CNDBGB_ratio = CDBGB/NDBGB

CPAGB_ratio = IF(PAGB>0.0001) THEN (CAGB/PAGB) ELSE(0)

CPBGB_ratio = IF(PBGB>0.0001)THEN(CBGB/PBGB)ELSE(0)

CPDAGB_ratio = CDAGB/PDAGB

CPDBGB_ratio = CDBGB/PDBGB

DON_conc_lake = 0.5

DON_conc_river = 1

DON_mineral_const = 0.0002

DONP_conc = IF(pore_water>0)THEN(DONP/pore_water)ELSE(0)

DONS_conc = IF(surface_water>0)THEN(DONS/surface_water)ELSE(0)

DOP_conc_lake = 0.1

DOP_conc_river = 0.1

DOP_mineral_const = 0.0002

DOPP_conc = IF(pore_water>0)THEN(DOPP/pore_water)ELSE(0)

DOPS_conc = IF(surface_water>0)THEN(DOPS/surface_water)ELSE(0)

evaporation_rate = GRAPH(COUNTER(0, 365))

(0.0, 5.245890411), (30.4166666667, 5.293150685), (60.8333333333, 5.245890411), (91.25, 5.529452055), (121.666666667, 4.773287671), (152.083333333, 6.164383562), (182.5, 5.136986301), (212.916666667, 4.109589041), (243.333333333, 5.136986301), (273.75, 5.623972603), (304.166666667, 5.907534247), (334.583333333, 4.300684932), (365.0, 5.245890411)

fraction_out = 0.01

growth_coeff_AGB = 0.3

growth_coeff_BGB = 0.2

harvest_%_of_AGB = 1

harvest_batch_yes = 0

harvest_day_batch = 230

harvest_day_regular = 1

harvest_in_g_AGB = 55

harvest_in_g_CAGB = harvest_in_g_AGB*C_conc_AGB_lit

harvest_interval_batch = 365

harvest_interval_regular = 1

harvest_regular_yes = 0

K_assim = 5150

K_denitri = 0.01

K_DON_diffusion = 0.1

K_DOP_diffusion = 0.1

K_NH4 = 0.7

K_NH4_diffusion = 0.1

K_nitri = 0.005

K_NO3 = 0.1

K_NO3_diffusion = 0.1

K_OP = 0.1

K_OP_diffusion = 0.1

K_radiance = 1

km_ads = 18.54

kmN = N_conc_papyrus_min+((N_conc_papyrus_optimal-N_conc_papyrus_min)/9)

kmP = P_conc_papyrus_min+((P_conc_papyrus_optimal-P_conc_papyrus_min)/9)

Kpor = 0.1

Ksurf = 0.1

Ktldown_C = 0.2

Ktldown_N = 0.2

Ktldown_P = 0.2

Ktlup_C = 0.2

Ktlup_N = 0.5

Ktlup_P = 0.5

lake_inflow_rate = 0.12

limit_N_ass = IF((N_conc_papyrus-N_conc_papyrus_min)/((kmN-N_conc_papyrus_min)+(N_conc_papyrus-N_conc_papyrus_min))<(0.9-N_conc_papyrus))AND((0.9-N_conc_papyrus)>N_conc_papyrus_min)AND(N_conc_papyrus>N_conc_papyrus_min)THEN((N_conc_papyrus-N_conc_papyrus_min)/((kmN-N_conc_papyrus_min)+(N_conc_papyrus-N_conc_papyrus_min)))ELSE IF((N_conc_papyrus-N_conc_papyrus_min)/((kmN-N_conc_papyrus_min)+(N_conc_papyrus-N_conc_papyrus_min))>=(0.9-N_conc_papyrus))AND((0.9-N_conc_papyrus)>N_conc_papyrus_min)AND(N_conc_papyrus>N_conc_papyrus_min)THEN(0.9)ELSE(0)

limit_NH4_uptake = NH4P_conc/(NH4P_conc+K_NH4)

limit_NO3_uptake = NO3P_conc/(NO3P_conc+K_NO3)

limit_NP_ass = (limit_N_ass*limit_P_ass)/0.81

limit_OP_uptake = OPP_conc/(OPP_conc+K_OP)

limit_P_ass = IF((P_conc_papyrus-P_conc_papyrus_min)/((kmP-P_conc_papyrus_min)+(P_conc_papyrus-P_conc_papyrus_min))<(0.9-P_conc_papyrus))AND((0.9-P_conc_papyrus)>P_conc_papyrus_min)AND(P_conc_papyrus>P_conc_papyrus_min) THEN(P_conc_papyrus-P_conc_papyrus_min)/((kmP-P_conc_papyrus_min)+(P_conc_papyrus-P_conc_papyrus_min))ELSE IF((P_conc_papyrus-P_conc_papyrus_min)/((kmP-P_conc_papyrus_min)+(P_conc_papyrus-P_conc_papyrus_min))>=(0.9-P_conc_papyrus))AND((0.9-P_conc_papyrus)>P_conc_papyrus_min)AND(P_conc_papyrus>P_conc_papyrus_min) THEN(0.9)ELSE(0)

limit_radiance = radiance/(radiance+K_radiance)

maint_coeff_AGB = 0.02

maint_coeff_BGB = 0.002

max_AGB_biomass = papyrus_max_biomass*AGB_total_biomass_ratio

max_assimilation_constant = 0.17

max_depth = 0.5

max_NH4_uptake = 0.1

max_NO3_uptake = 0.1

max_OP_uptake = 0.1

mode = IF(water_filled_porosity>1) OR(water_filled_porosity=1) THEN(0) ELSE IF(water_filled_porosity>water_filled_porosity_field_capacity) AND(water_filled_porosity<1) THEN((1-water_filled_porosity)/(1-water_filled_porosity_field_capacity)) ELSE(1)

N_AGB_to_BGB_optimal_ratio = NAGBlitavg/NBGBlitavg

N_conc_AGB = IF (AGB>0) THEN(NAGB/AGB) ELSE (0)

N_conc_AGB_lit = NAGBlitavg/AGBlitavg

N_conc_BGB = IF(BGB>0)THEN(NBGB/BGB)ELSE(0)

N_conc_BGB_lit = NBGBlitavg/BGBlitavg

N_conc_papyrus = IF(C_papyrus>0)THEN((N_papyrus/C_papyrus)*C_conc_papyrus)ELSE(0)

N_conc_papyrus_min = 0.0016

N_conc_papyrus_optimal = N_max_papyrus/papyrus_max_biomass

N_initial_DAGB = papyrus_initial_dead_biomass*AGB_total_biomass_ratio*N_conc_AGB_lit

N_initial_DBGB = papyrus_initial_dead_biomass*BGB_total_biomass_ratio*N_conc_BGB_lit

N_max_AGB = papyrus_max_biomass*N_conc_AGB_lit*AGB_total_biomass_ratio

N_max_BGB = papyrus_max_biomass*N_conc_BGB_lit*BGB_total_biomass_ratio

N_max_papyrus = N_max_AGB+N_max_BGB
N_papyrus = NAGB+NBGB
N_retrans_constant = 0.7
NAGBlitavg = 44
NBGBlitavg = 31
NH4_conc_lake = 1
NH4_conc_river = 3
NH4P_conc = IF(pore_water>0)THEN(NH4P/pore_water)ELSE(0)
NH4S_conc = IF(surface_water>0)THEN(NH4S/surface_water)ELSE(0)
NO3_conc_lake = 1
NO3_conc_river = 2
NO3P_conc = IF(pore_water>0)THEN(NO3P/pore_water)ELSE(0)
NO3S_conc = IF(surface_water>0)THEN(NO3S/surface_water)ELSE(0)
OP_conc_lake = 0.3
OP_conc_river = 0.5
OPADS_eq = OPADS_max*OPP_conc/(OPP_conc+km_ads)
OPADS_max = OPADS_maxdw*bulkdens*soil_depth
OPADS_maxdw = 0.004
OPP_conc = IF(pore_water>0)THEN(OPP/pore_water)ELSE(0)
OPS_conc = IF(surface_water>0)THEN(OPS/surface_water)ELSE(0)
P_AGB_to_BGB_optimal_ratio = PAGBlitavg/PBGBlitavg
P_conc_AGB = IF(AGB>0)THEN(PAGB/AGB)ELSE(0)
P_conc_AGB_lit = PAGBlitavg/AGBlitavg
P_conc_BGB = IF(BGB>0)THEN(PBGB/BGB)ELSE(0)
P_conc_BGB_lit = PBGBlitavg/BGBlitavg
P_conc_papyrus = IF(C_papyrus>0)THEN((P_papyrus/C_papyrus)*C_conc_papyrus)ELSE(0)
P_conc_papyrus_min = 8e-05
P_conc_papyrus_optimal = P_max_papyrus/papyrus_max_biomass
P_initial_DAGB = papyrus_initial_dead_biomass*P_conc_AGB_lit*AGB_total_biomass_ratio
P_initial_DBGB = papyrus_initial_dead_biomass*P_conc_BGB_lit*BGB_total_biomass_ratio
P_max_AGB = papyrus_max_biomass*P_conc_AGB_lit*AGB_total_biomass_ratio
P_max_BGB = papyrus_max_biomass*P_conc_BGB_lit*BGB_total_biomass_ratio
P_max_papyrus = P_max_AGB+P_max_BGB
P_papyrus = PAGB+PBGB
P_retrans_constant = 0.77
PAGBlitavg = 2.61
papyrus_biomass = BGB+AGB
papyrus_biomass_litavg = AGBlitavg+BGBlitavg
papyrus_initial_dead_biomass = 1340
papyrus_max_biomass = 8118
PBGBlitavg = 2.78
POC_hydrolysis_constant = 0.0001
POC_settling_rate = 0.05
PON_conc_lake = 0.5
PON_conc_river = 1
PON_hydrolysis_const = 0.00063
PON_settling_constant = 0.05
PONP_conc = IF(pore_water>0)THEN(PONP/pore_water)ELSE(0)
PONS_conc = IF(surface_water>0)THEN(PONS/surface_water)ELSE(0)
POP_conc_lake = 0.1
POP_conc_river = 0.1
POP_hydrolysis_constant = 0.00063
POP_settling_constant = 0.05
POPP_conc = IF(pore_water>0)THEN(POPP/pore_water)ELSE(0)
POPS_conc = IF(surface_water>0)THEN(POPS/surface_water)ELSE(0)
pore_water = Water-surface_water

porewater_free = IF(water_fraction>porosity) THEN((1-water_filled_porosity_field_capacity)*porosity*soil_volume) ELSE IF(water_fraction>water_filled_porosity_field_capacity*porosity) AND(water_fraction<porosity) OR(water_fraction=porosity) THEN((water_fraction-water_filled_porosity_field_capacity*porosity)*soil_volume) ELSE(0)

porosity = 0.8

radiance = GRAPH(COUNTER (0, 365))

(0.0, 23.500), (30.4166666667, 23.500), (60.8333333333, 23.000), (91.25, 23.500), (121.666666667, 21.000), (152.083333333, 20.500), (182.5, 20.500), (212.916666667, 19.000), (243.333333333, 20.500), (273.75, 22.000), (304.166666667, 22.000), (334.583333333, 21.000), (365.0, 22.500)

rainfall_rate = GRAPH(COUNTER (0, 365))

(0.0, 0.118356164), (30.4166666667, 0.147945205), (60.8333333333, 0.328767123), (91.25, 1.97260274), (121.666666667, 6.246575342), (152.083333333, 2.95890411), (182.5, 2.465753425), (212.916666667, 2.465753425), (243.333333333, 1.315068493), (273.75, 1.315068493), (304.166666667, 0.821917808), (334.583333333, 1.315068493), (365.0, 0.118356164)

river_inflow_rate = GRAPH(COUNTER (0, 365))

(0.0, 0.001), (30.4166666667, 0.056145834), (60.8333333333, 0.125416668), (91.25, 0.129791668), (121.666666667, 0.160416668), (152.083333333, 0.135), (182.5, 0.1125), (212.916666667, 0.0005), (243.333333333, 0.0001), (273.75, 0.0002), (304.166666667, 0.0025), (334.583333333, 0.0002), (365.0, 0.001)

soil_depth = 0.2

soil_volume = soil_depth*1

surface_water = IF(water_fraction<porosity) OR(water_fraction=porosity) THEN(0) ELSE((water_fraction-porosity)*soil_volume)

tldownmax_C = 15

tldownmax_N = 0.5

tldownmax_P = 0.5

tlupmax_C = 0.5

tlupmax_N = 5

tlupmax_P = 1

water_filled_porosity = water_fraction/porosity

water_filled_porosity_field_capacity = 0.625

water_fraction = Water/soil_volume

wet_yes_or_no = 0

{ The model has 317 (317) variables (array expansion in parens).

 In root model and 0 additional modules with 4 sectors.

 Stocks: 30 (30) Flows: 104 (104) Converters: 183 (183)

 Constants: 94 (94) Equations: 193 (193) Graphicals: 4 (4)

 }

6

THE EFFECT OF HARVESTING AND FLOODING ON NUTRIENT CYCLING AND RETENTION IN *CYPERUS PAPYRUS* WETLANDS – SYNTHESIS, CONCLUSIONS AND RECOMMENDATIONS

6.1 SYNTHESIS

Wetlands are often considered to be natural water treatment systems within the landscape because they store, immobilize and remove sediments and nutrients. Quantification of these ecosystem functions can contribute to improving landscape and wetland management and to determining trade-offs between provisioning and regulating ecosystem services. To quantify these regulating ecosystem services, it is important to address questions like: What are the mechanisms behind these ecosystem functions and services? What happens to specific nutrients like nitrogen (N) and phosphorus (P), are they permanently removed or temporarily stored and what are the main processes involved? What happens to the regulating services if the wetland is under pressure from anthropogenic (e.g. conversion to agriculture) or natural drivers (e.g. fluctuating water levels)? This thesis looks into these questions for wetlands dominated by the sedge, *Cyperus papyrus* (L.), commonly referred to as papyrus wetlands. Papyrus wetlands occur throughout eastern and southern Africa and the Middle East, and support the livelihoods of millions of people through food provisioning and other important ecosystem services, including the retention of nutrients (Kipkemboi and van Dam 2018).

The overall objective of this thesis was to develop a dynamic simulation model for nutrient retention in papyrus wetlands to support the analysis of trade-offs between provisioning ecosystem services and regulating ecosystem services, particularly N and P retention. Chapter 1 introduced this objective, presented an overview of the literature on papyrus wetlands, and explained the structure of the thesis chapters with their specific objectives. Chapter 2 then presented the results of field experiments in two papyrus wetlands in East Africa that focused on the role of living aboveground biomass in the uptake and storage of N and P under different degrees of human disturbance and varying flooding regimes. Chapter 3 presented the literature on wetland modelling and a conceptual model for quantifying N and P retention, and identified components of existing wetland models that could be useful for developing a papyrus wetland model. Chapter 4 then introduced the development and application of a new papyrus model, the 'Papyrus Simulator', with a description of N retention at the process level and how this was affected by wetland hydrology and by vegetation harvesting. The model was parameterized with data from the literature and from field experiments, and was able to predict the biomass of papyrus, the nutrient concentrations in the soil and surface water, and the retention of N and P under different harvesting regimes. In Chapter 5, a second version of the Papyrus Simulator was introduced with an improved hydrology sub-model and an additional sub-model for P cycling and retention. Chapter 5 also included a sensitivity analysis, simulation results on N and P retention and N:P output ratios under different harvesting regimes and hydrological conditions.

The role of aboveground biomass of papyrus in the storage and retention of N and P was studied in two wetland sites in Kenya and Tanzania under seasonally and permanently flooded conditions (Chapter 2). The field experiments had a duration of 3 months and samples were collected along transects perpendicular to open water in a river. Each transect was dominated by papyrus vegetation and included a seasonally flooded zone away from the river, and a permanently flooded zone closer to the open water. The first site, Nyando wetland (Kenya) was under anthropogenic disturbance from agriculture and vegetation harvesting, whereas the second site, Mara wetland (Tanzania), was less disturbed. Maximum papyrus culm growth in both sites was described well by a logistic model (regressions for culm length with R^2 from 0.70 to 0.99), with culms growing faster but less tall in Nyando compared with Mara. In both sites, the maximum culm length was greater in permanently than in seasonally flooded zones, young shoots had higher N and P concentrations in their biomass than mature shoots, and the highest amount of N and P in a single culm occurred before the maximum length was reached. Both the total aboveground biomass and the amounts of N and P stored per unit area were higher in Mara than in Nyando. In disturbed sites (Nyando), papyrus plants showed characteristics of r-selected species, with faster growth but lower biomass and nutrient storage than the plants with K-selected characteristics in the more pristine sites (Mara). These findings increase the understanding of N and P storage in papyrus biomass, and enable quantification of impacts of livelihoods activities and inundation on N and P retention in natural wetlands. More disturbance by harvesting and seasonal agriculture means more re-growth of aboveground biomass and higher uptake of nutrients from the water. On the other hand, it also means that less nutrients are stored per unit area. On a catchment scale, what happens with the harvested papyrus (e.g. use within the catchment, or export from the catchment) determines if harvesting leads to a removal of the nutrients or a release back into the system. Besides a better understanding of natural papyrus wetlands, the results support optimization of harvesting regimes for aboveground papyrus biomass in constructed wetlands to increase removal of N and P from wastewater. As the highest amount of both N and P in the aboveground biomass is reached before maximum height, timely harvesting will improve the nutrient removal efficiency of a constructed wetland.

A review of existing wetland models revealed the modelling requirements to quantify N and P retention under pressure from different water levels and harvesting intensity (Chapter 3). Analysing existing models resulted in an overview with four categories: a) hydrological models; b) biogeochemical models; c) vegetation models; and d) integrated models. A more narrow focus on the processes underlying N and P retention required a more detailed analysis of the models in the biogeochemical (b) and vegetation (c) categories. All models within these categories that specifically dealt with retention were selected. These 10 models provided insights on how to model relevant processes such as uptake of N and P, plant growth, mortality, decay, P adsorption, nitrification and denitrification. Of special importance were the models on Kismeldon Meadows (van der Peijl and Verhoeven 1999; 2000), a floating papyrus model (van Dam et al. 2007), a collection of general macrophyte models (Asaeda and Karunaratne 2000; Asaeda et al. 2008; 2011), and the marsh version of a shallow lake model, PC Lake (Janse et al. 2001; Sollie et al. 2008). The proposed concept of a papyrus wetland model (Figure 3.1, Chapter 3) showed the impact of hydrological variations and harvesting on water quality outputs, and N and P retention. The conceptual model illustrated the possibility of a quantified comparison of the contribution of different processes (e.g. peat formation, adsorption, uptake, denitrification) to N and P retention in papyrus wetlands.

To understand the processes contributing to N retention and to evaluate the effects of papyrus harvesting, a dynamic model for N cycling in rooted papyrus wetlands, called Papyrus Simulator, was constructed (Chapter 4). The hydrological sub-model simulated seasonally flooded zones and permanently flooded zones and was based on data from papyrus wetlands fringing Lake Naivasha, Kenya. A carbon sub-model described carbon (C) assimilation and respiration to simulate the growth of papyrus vegetation. In each zone, the flows of water, N and C were calculated based on descriptions of hydrological (river flow, lake level, precipitation, evaporation) and ecological (e.g. photosynthesis, N uptake, mineralisation, nitrification) processes. The Papyrus Simulator was than expanded with a P sub-model (Chapter 5). Literature data were used for parameterization and calibration. A comparison with published data on 43 papyrus wetland studies showed that the model simulated realistic papyrus biomass and concentrations of dissolved N and P in the water. The model outputs showed that the relative extent of nutrient retention reduced the N:P ratios in the water column to around 20 N:P molar ratio. The seasonal absence of surface water in the dry season caused a temporary reduction of papyrus biomass, due to nutrient limitation as a result of reduced input through surface water. Harvesting increased N retention from 7% to over 40%, and P retention from 4% to 40%, due to the increased uptake of N and P by regrowth of biomass. Sensitivity analysis revealed that assimilation, mortality, decay, re-translocation, nutrient inflow and soil porosity were the most influential factors for retention. It was concluded that the Papyrus Simulator is suitable for quantifying nutrient retention and N and P concentrations in the outflowing water through quantification of the underlying processes, and that the model can evaluate the effects of papyrus biomass harvesting and varying water levels on N and P processes in papyrus wetlands.

The results presented in this thesis can be used for local management of papyrus wetlands in balancing between the regulating service of water purification and various provisioning services such as seasonal agriculture and direct use of papyrus biomass. Papyrus wetlands are found around major lakes and river floodplains in East Africa, such as the Lake Victoria basin. These inland waters suffer increasingly from eutrophication and related issues such as changes in fish communities and proliferation of water hyacinth and cyanobacteria (Olokotum et al. 2020). Quantifying the contribution of papyrus wetlands to avoiding N and P runoff into these lakes enables local natural resource managers and policy makers to mitigate the negative economic impacts of eutrophication on society. The role of wetlands in regional and global models is often neglected or simplified, yet ecological feedbacks in modelling of earth systems is deemed important (Bonan and Doney 2018). Explanatory models like the Papyrus Simulator can therefore help to improve regional and global modelling efforts.

6.2 NITROGEN AND PHOSPHORUS RETENTION IN WETLANDS

Wetlands are considered as natural treatment systems, especially for N and P. Biomass plays an important role in nutrient retention, as especially large macrophytes can contain high amounts of N and P. Due to the anoxic conditions in wetlands, dead biomass will not fully decompose and therefore accumulate. *Cyperus papyrus* plants can grow fast and contain large amounts of N (75 g N m^{-2}) and P (5.5 g P m^{-2}) (Chapters 1; 5). Understanding how N and P in living biomass varies over time and responds to management regimes (harvesting) or different inundation levels will enable further quantification of N and P retention in papyrus wetlands.

The field experiments showed that papyrus growth can be described with a logistic equation, growing to a maximum height in 5-6 months. Under permanently flooded conditions and with relatively little impact of human livelihoods activities the papyrus grows tallest (5 m). With seasonally flooded conditions and under higher impact of human activities, papyrus grows faster, but attains a lesser height (4 m). Also the biomass (7 kg m^{-2}, compared with 2 kg m^{-2}) and culm density (25 culms m^{-2}, compared with 20) were higher under less disturbed conditions. The concentration of N and P stored in culms was higher when they started developing than in mature culms. The absolute amounts of N and P in one culm were highest before they reached maximum height and maturity (Chapter 2).

The use of papyrus (e.g. harvesting for direct use or seasonal agriculture) can decrease N and P concentrations in surface water because papyrus culms grow back fast after harvesting and then incorporate large amounts of N and P. This characteristic can also be used to actively manage and remove N and P from the system by planned harvesting. Constructed or modified natural papyrus wetlands are seen as 'nature based solutions' to reduce N and P concentrations in surface water (Haddis et al. 2020). The frequency of harvesting aboveground biomass to optimize removal efficiency in constructed wetlands can be improved based on findings in Chapter 2. On the other hand frequent harvesting may lead to a less healthy stand of papyrus (Terer et al. 2012a), reduce the accumulation of organic matter and expose existing organic matter to more oxygen rich conditions, increasing decomposition and releasing stored N and P.

6.3 WETLAND MODELLING

Wetlands are largely absent in earth system models (Bonan and Doney 2018), especially in applications that go beyond research (FAO/IWMI 2018). Modelling of freshwater ecosystems to improve management and policy development for use and conservation is, however developing continuously (Mooij et al. 2019). One of the challenges is the modelling of nutrient cycling and water quality in wetland ecosystems (Janssen et al. 2015). Despite the presence of a range of research models at the local wetland site level, especially for wetlands in temperate regions, the impact of wetlands on water quality in other (climate) regions and upscaling to regional or even global scale remains a challenge (Chapter 3). FAO has suggested the use of riparian buffer zones and constructed wetlands to mitigate the impact of agriculture on water quality, and identified the need for ecological models applicable for policy development (FAO/IWMI 2018). The IPCC has recognized the importance of wetlands in capturing and emitting greenhouse gasses and suggested modelling as a method for national greenhouse gas inventories from wetlands (Hiraishi et al. 2013). At the same time we are still far from understanding the impacts of climate change on wetlands, its biodiversity, water levels and loss of stored carbon and other nutrients (Finlayson 2018; Moomaw et al. 2018). This further emphasizes the need for explanatory wetland models for research and for policy development.

To address the need for wetland models in general and for tropical regions in particular, developing a papyrus wetland model is highly relevant. Papyrus wetlands still cover a large area (estimated between 20,000 – 85,000 km^2) and have a high societal relevance (Kipkemboi and van Dam 2018). The main independent variables of the Papyrus Simulator developed in this thesis were the hydrological processes (inflow; evapotranspiration; precipitation; seepage) and environmental factors (light; N and P loading). The Papyrus Simulator is a "square meter model" with a simple hydrological model as a forcing factor, which makes it generally applicable at local, regional and global levels. The alternative would have been to develop a model for a unique papyrus wetland system with its specific hydrology as a forcing factor, which would make the model only applicable to that specific wetland site. The advantages of a site-specific model could be more accurate predictions of system characteristics at the site level. By connecting the Papyrus Simulator to a local hydrological model (including N and P loading rates), N and P retention for a specific area can be estimated and the effect of the wetland on outflowing N and P concentrations as well as the outgoing N:P ratio. This was done for C, N and water dynamics in a Cypress Wetland-Pine Upland ecosystem by coupling the Wetland-DNDC model to the spatially explicit MIKE SHE hydrological model (Sun et al. 2006). The Papyrus Simulator can function as a grid cell driven by hydrology in a dynamic spatial model. By varying the area and introducing different harvesting regimes, the impact of different scenarios (conversion; sustainable use; conservation) can be compared. Scenarios for different ratios between intact and converted wetland and moderately and intensively used area could be assessed on water quality impact. The model can compare the contribution of different processes (e.g. organic matter accumulation, denitrification, uptake, adsorption) to N and P retention and how this is affected by harvesting and hydrology (Chapters 4 and 5). By coupling this model to existing regional or global hydrological models, N and P loading and climate data, N and P retention in specific wetlands (e.g. Lake Victoria's papyrus wetlands) can be quantified or form part of a global assessment. In fact, Papyrus Simulator is one of the prototype models for the development of a global model to quantify wetland ecosystem services which is coupled to the global hydrological model PCR-GLOBWB (Janse et al. 2019; see below).

6.4 WETLAND ECOSYSTEM SERVICES TRADE-OFFS

Wetland conversion is often a result of the expansion of agricultural land into these fertile and productive areas, leading to an increase of food production locally. However on a larger scale, conversion also leads to loss of regulating services and biodiversity that will negatively impact water availability for agriculture and water quality. Locally, there is also loss of livelihoods support, through lower availability of fish, fiber and seasonal agriculture in the natural wetland. The suite of benefits that wetlands provide (e.g. water purification, biodiversity hotpots, local climate regulation, water storage, carbon storage, fish nursery) makes them essential, both as components of landscape ecology and for human well-being. The total value of wetlands is estimated at Int\$47.4 trillion per year, which is 43.5 % of all natural biomes, while only covering 3% of the global surface area. Of this total value, Int\$14.5 trillion per year is attributed to swamps and floodplains such as papyrus wetlands (Davidson et al. 2019). For inland wetlands the estimated value of nutrient cycling alone was estimated higher (1,713 Int\$ ha^{-1} yr^{-1}) than all combined provisioning services like food and water (1,659 Int\$ ha^{-1} yr^{-1}), and the total value of all regulating, habitat, and cultural ecosystem services was estimated at 24,022 Int\$ ha^{-1} yr^{-1}, which is 14 times higher than the value of the provisioning services (De Groot et al. 2012). Still, the focus in management is often biased towards provisioning services as their values are associated with direct economic benefits. This leads to underestimating the more long-term and more valuable regulating services (Costanza et al. 2017; Balasubramanian 2019). Quantifying these benefits in biophysical units is essential for estimating their economic values and needed to make the trade-offs between provisioning and regulating services more visible and explicit. This will help wetland managers and policy development to better combine short-term with long-term interests.

One of the central questions in the conservation of natural ecosystems and biodiversity is: should we designate and protect high-biodiversity wetland areas to optimize regulating services, cultural services, and habitat services, and sacrifice other wetlands to increase (intensive) food production and other provisioning services ('sparing')? Or, should wetland management focus on a balanced use of all benefits in wetlands generally ('sharing')? Kremen (2015) concluded in her review that the choice between sparing and sharing is too rigid. A choice may increase fragmentation ('sparing'), while connectivity between natural areas is essential for biodiversity conservation, or to loss of fragile habitat and biodiversity ('sharing'). She therefore suggested protected areas surrounded by wildlife-friendly farming matrices, and calls for increased research efforts into suitable farming methods that are wildlife friendly, but also sufficient to feed the growing population. For the management of East African papyrus wetlands the choice between converting, sustainable use or conserving is highly relevant and inevitably people-centered because so many people are directly dependent on wetlands (Wood 2013). Food demands are rising and the technology and knowledge to drain and convert wetlands are available. However, wetlands are also essential in ensuring 'green water' (soil moisture from rainfall) availability on which 95% of sub-Saharan African agriculture depends (Rockström and Falkenmark 2015). Wetlands dampen flash floods, and maintain underground flows to help avoid erosion, and to keep green water in the system for food production (Savenije 1995; Gordon et al. 2010; Rockström and Falkenmark 2015). Moreover water quality in the larger African lakes that are important for fisheries is deteriorating, with eutrophication impacting fish diversity and fish stocks, on top of the impact of the fisheries sector itself (Irvine et al. 2018; Olokotum et al. 2020). Further downward trends in the area and quality of inland wetlands will exacerbate the nutrient enrichment of lakes and coastal zones in Africa. Wetlands are also important as nurseries for a variety of fish species, maintaining diversity (Hickley et al. 2004; Koning et al. 2020). Summarizing, the conversion of wetlands will have disastrous consequences for biodiversity, water quality, food production and overall human well-being.

The Papyrus Simulator can quantify the impacts on N and P retention and surface water concentrations for scenarios with different areas of protected, moderately used or converted papyrus wetland, as explored in the previous paragraph. The impact of agriculture or any other land use change itself is currently not included in the model, however other models are available to estimate N and P balances for specific crops (Kollas et al. 2015; Gallardo et al. 2020). Papyrus Simulator also does not quantify other ecosystem services that the wetlands provide (e.g. biodiversity habitat, food and water, climate regulation, fish nursery), but there are tools available that quantify a wide range of ecosystem services both biophysically and economically (Kotze et al. 2009; Maltby 2009; Peh et al. 2013; Villa et al. 2014). A future version of the model could integrate other land uses and a wider range of ecosystem services, and ultimately develop towards a social-ecological model for transdisciplinary research and participatory decision making (Pohl and Hadorn 2008).

6.5 MANAGEMENT APPLICATIONS AND POLICY CONTRIBUTIONS

For papyrus wetlands, the application of the Papyrus Simulator as well as the results from the field experiments can support policy development and management, and be a starting point for transdisciplinary social-ecological research. Legitimacy in wetland governance is influenced by legal responsibilities, leadership, resources, public support and knowledge (Guzman et al. 2011). The quantification of an important ecosystem function, such as N and P retention, could empower local decision makers and shift governance from a more hierarchical mode, with a strong role for the regional or local government, to community governance with more influence of community-based organization (CBOs) or informal leadership to protect the ecosystem on which the community depends. Besides CBOs, also national and international NGOs can be empowered by providing quantitative data and jointly developed argumentation to support wetland management. For example, in the on-going decision making process related to the installation of dams in the lower Mara River in Tanzania, which creates a threat to the existence of the Mara Wetland (Mnaya et al. 2017), quantitative data on the regulating ecosystem services of the wetland can provide arguments to develop more sustainable management regimes. The model can also support a more evidence-based adaptive management approach, by including local perceptions and ecological knowledge in managing papyrus wetlands (Terer et al., 2012b; Morrison et al., 2013) and transdisciplinary research. The model could provide input to Bayesian belief networks, which are models that can combine stakeholder perceptions and perspectives with biophysical data. This was demonstrated by linking hydrology, ecosystem function, and livelihood outcomes in Nyando wetland in Kenya (van Dam et al. 2013). Another example is a transdisciplinary approach to manage groundwater contamination in Denmark, where opposing views of hydrologists and farmers were modelled to help negotiations with transparency and understanding the different perspectives (Henriksen et al. 2007).

The results of this research were incorporated into two educational modules on environmental modelling at IHE Delft and frequently used by students in a module on environmental systems analysis. Outside IHE the results were used annually in a module on wetland management at Centre for Development Innovation, Wageningen University and Research, and for the past two years at the Autonomous University of Mexico (UNAM) in Mexico City. A total of 10 MSc students have used or contributed to this research through their MSc thesis project and are now using this knowledge in their respective countries: Bhutan, China, Colombia, Ethiopia, Kenya(2), Tanzania, Uganda(2) and Vietnam.

This thesis contributes to achieving five SDGs directly (SDG2 zero hunger; SDG6 clean water and sanitation; SDG13 climate action; SDG14 life below water; and SDG15 life on land) and indirectly contributes to achieving SDG 1 (no poverty) and SDG 10 (reduced inequality). Figure 6.1 illustrates the ecosystem services that are provided by papyrus wetlands, and the underlying ecosystem functions that can be quantified by the Papyrus Simulator. When linked to a spatially explicit hydrological model, the Papyrus Simulator can produce information on water residence times, inundation levels, N and P fluxes and above- and belowground biomass. The model can be further developed and expanded with a more elaborate carbon section and quantify carbon fluxes. These outputs can be used to quantify ecosystem functions (Maltby 2009) as proxies of provisioning services (food; water), regulating services (water purification; local climate regulation; moderation of extreme events; carbon sequestration and storage), and habitat services (habitats for species). These ecosystem services contribute directly to the five SDGs in the categories food, water, climate and biodiversity (Ramsar Convention 2018), but also indirectly to poverty alleviation (SDG1), as millions depend on papyrus wetlands for their livelihoods (Kipkemboi and van Dam 2018); and reduced inequality (SDG 10), as healthy wetlands mitigate the risk of poor access to water (Ramsar Convention 2018).

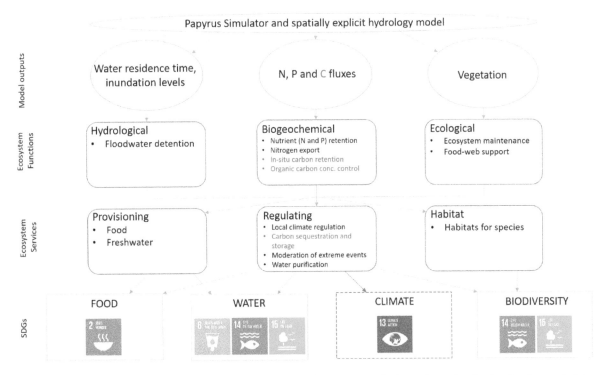

Figure 6.1 Contributions of the Papyrus Simulator coupled with a hydrological model to achieving the Sustainable Development Goals through model outputs, ecosystem functions and ecosystem services. In red elements that assume an incorporated carbon model. Modified from Janse et al. (2019)

The development of a global wetland model (Janse et al. 2019), that was partly based on this thesis, has the potential to quantify the role of wetlands in global fluxes of C, N and P, water, and vegetation biomass and to reveal the impact of wetland loss on ecosystem services. The global model will help filling gaps in current estimations of carbon fluxes for climate modelling (Bonan and Doney 2018; Finlayson 2018). It can show the impact of current wetland loss rates (Davidson et al. 2020) on water quality and the availability of clean drinking water. Moreover, with vegetation cover as a proxy, a global model can illustrate the impact of area reduction on biodiversity loss, as wetlands are highly diverse ecosystems, supporting the life of terrestrial, aquatic, and specialized wetland species. The impact of wetland decline on local and regional water cycles would be a good indicator for reduction in food production potential (rainfed and irrigated agriculture; wetland fish and fruits; extensive wetland agriculture). Quantifying and communicating the threats to human well-being caused by the rapid decline of ecosystem services associated with wetlands and other biomes is crucial to generate global awareness and the political will to overcome the immense challenges towards a sustainable future (Finlayson et al. 2019; Bradshaw et al. 2021).

6.6 OPPORTUNITIES FOR FURTHER RESEARCH

The field experiments were focused on aboveground biomass and less on belowground biomass and peat. However, based on the modelling work, the accumulation of organic matter (ultimately leading to peat formation) was the main mechanism for retention (Chapters 4 and 5). Follow-up experiments could therefore focus on belowground biomass and especially on peat formation. Relatively little is known about peat layers in papyrus wetlands, and even in recent studies the information is mostly based on estimates or measurements from 40 years ago rather than recent empirical work (Saunders et al. 2014). This is partly due to the inaccessibility of especially the larger papyrus wetlands such as the Sudd in South Sudan and the Lake Tumba-Lidiima Reserve in DR Congo. Information that is available shows a large variation in depth of peat layers, from 1.5 to 10 meters (Saunders et al. 2014). While assuming a modest 1 m average peat depth, an estimated 640 t C ha^{-1} is stored (Jones and Humphries 2002), which is more than in tropical rainforests. It would be interesting to further investigate what these peat reserves mean for N and P storage and release, and to what extent loss and degradation of papyrus wetlands in Africa contribute to greenhouse gas emissions. Potential denitrification assays in the Namatala wetland in eastern Uganda, both in intact papyrus stands and in other parts of the wetland that were converted to agriculture, showed that conversion and drainage of the wetland significantly reduces the C content of wetland soils and can increase the production of N_2O (Gettel et al. 2019; Namaalwa et al. 2020).

Chapter 5 presented an overview of the literature on papyrus biomass, with a range of values for biomass density of aboveground and belowground biomass and for floating as well as rooted papyrus stands. The analysis could not explain the differences in biomass from environmental conditions related to climate or altitude. The two field sites in this research had roughly the same climate conditions and were at the same altitude. To get more insight in the role of altitude and climate on biomass densities, field experiments comparing biometric parameters at different altitudes and climate regions could be an interesting follow up of the work in this thesis. Findings could be used to include climate parameters (e.g. air and water temperature) and altitude in the Papyrus Simulator and widen the applicability of the model.

Chapters 4 and 5 list reasons for scoping out or simplifying some of the processes and forcing factors related to the N and P cycles. However, there were also important processes that were not included in the current model: biological N fixation, atmospheric deposition of N and P and sediment retention. The role of N fixation from the atmosphere as well as from microbial N fixation in the root system is largely unknown in papyrus wetlands. The limited published research shows that N fixation in the young roots alone could supply over 25% of the N requirements for papyrus plant growth (Mwaura and Widdowson 1992). Dry and wet deposition of N and P do occur, but the amount is relatively small (estimated 710 mg N m^{-2} yr^{-1}), although this may change in the future as a result of both industrial and agricultural developments (Dentener et al. 2006). Sediment retention is an important ecosystem service of papyrus wetlands (Boar and Harper 2002; Bregoli et al. 2019), this was partly included in the Papyrus Simulator (particulate organic matter), however an expansion of the model is recommended to better quantify mineral and organic sediment retention. Oxygen was included in the Papyrus Simulator and is important for a range of wetland processes, but in the current model only influences nitrification and denitrification, and using a rather simplified process description. Temperature dependence of the processes in the Papyrus Simulator was not explicitly modelled, which is acceptable for ecosystems located in the proximity of the equator which have a relatively stable temperature regime throughout the year. For a wider applicability and for systems that have seasonal variation in water and air temperatures, temperature-dependent processes will need to be included in the model, as was shown in a preliminary study with the Papyrus Simulator (Osorio 2018). Process descriptions that include oxygen, temperature and other factors like redox potential are available from the literature (Chapter 3). However, there is insufficient data to use these more complex descriptions with confidence.

Absence of surface water inflows reduces transport and availability of nutrients and eventually nutrient limitation. Water itself can also be a limiting factor for plant growth. The Papyrus Simulator only includes the effect of nutrient limitation when water is no longer available. Plant models such as the WOFOST (WOrld FOod STudies) field crop model do include process descriptions for water limitation (de Wit et al. 2019) and could be used as a starting point to include water limitation in the papyrus model. While it is not clear how this process affects papyrus vegetation, it is possible that papyrus culms die faster in the absence of water then is currently modelled.

Modelling of data scarce systems is a challenge for parameterization, calibration and validation. For the papyrus model, parameterization and calibration was done primarily using literature data related to east African papyrus wetlands, notably studies of Lake Naivasha, but also other papyrus wetlands or even entirely other types of wetland systems. Although validation could not be done with time series from an independent dataset, it was possible to compare model outputs with independent measurements from the literature (Chapter 5). As long as regular monitoring of environmental parameters is limited and specific measurements of vegetation and processes related to N and P retention are scarce, conventional model validation will remain problematic. However, alternative non-invasive monitoring options are developing rapidly. Remote sensing with satellite imagery is now able to distinguish different types of vegetation through machine learning and artificial intelligence (Mutanga et al. 2012; Zhu et al. 2017; Xu et al. 2020). Very soon, monitoring of current biomass as well as analysing historical imagery to create time series of wetland ecosystem development will become promising validation options. Even more detailed images can be obtained by unmanned aerial vehicles (UAVs or drones), which allows monitoring of individual culm growth and detailed observation of pressures like harvesting, burning and water level fluctuations (Bregoli et al. 2019).

Another validation option is to have controlled experiments with constructed wetlands and generate datasets that cover a period of at least 2 years and different harvesting scenarios. This will allow validation of the whole development cycle of the papyrus culms in terms of N and P content, biomass, and density and how this is impacted by multiple regrowth periods after harvest. Having a validated model for constructed wetlands would also increase to applicability of the Papyrus Simulator to design and operation of constructed wetlands and natural systems as nature-based solutions (FAO-IWMI 2018; Haddis et al. 2020).

A final recommendation is to introduce a more complete carbon section to the Papyrus Simulator to quantify contributions to greenhouse gas (GHG) fluxes. This could be done by collecting data on carbon fluxes from the literature and using the Denitrification-Decomposition model (DNDC) equations (EOS, 2017) as starting point. The current carbon section was designed to model carbon assimilation, which is sufficient for simulating papyrus vegetation growth, but not for calculating overall carbon fluxes, including the greenhouse gases CO_2 and CH_4. A full carbon model, and including N_2O emissions in the nitrogen section would allow the quantification of GHGs and widen the scope and applicability.

7
REFERENCES

Abira MA, van Bruggen JJA, Denny P (2005) Potential of a tropical subsurface constructed wetland to remove phenol from pre-treated pulp and papermill wastewater. Water Science and Technology 51(9):173-176.

Adam E, Mutanga O, Abdel-Rahman EM, Ismail R (2014) Estimating standing biomass in papyrus (*Cyperus papyrus* L.) swamp: Exploratory of in situ hyperspectral indices and random forest regression. Int J Remote Sens 35(2):693-714

Ajwang' Ondiek RA, Kitaka N, Oduor SO (2016) Assessment of provisioning and cultural ecosystem services in natural wetlands and rice fields in Kano floodplain, Kenya. Ecosystem Services 21:166-173

APHA (1992) Standard methods for examination of water and wastewater. American Public Health Association, Washington DC, USA

Asaeda T, Baniya MB, Rashid MdH (2011) Effect of floods on the growth of *Phragmites japonica* on the sediment bar of regulated rivers: a modelling approach. International Journal of River Basin Management 9(3-4):211-220

Asaeda T, Hai DN, Manatunge J, Williams D, Roberts J (2005) Latitudinal Characteristics of Below- and Above-ground Biomass of *Typha*: a Modelling Approach. Annals of Botany 96(2):299-312

Asaeda T, Karunaratne S (2000) Dynamic modelling of the growth of *Phragmites australis*: model description. Aquat Bot. 67(4):301-318

Asaeda T, Rajapakse L, Fujino T (2008) Applications of organ-specific growth models; modelling of resource translocation and the role of emergent aquatic plants in element cycles. Ecol Model 215(1-3):170-179

Augusiak J, van den Brink PJ, Grimm V (2014) Merging validation and evaluation of ecological models to 'evaludation': A review of terminology and a practical approach. Ecol Model 280:117-128

Azza NGT, Denny P, van de Koppel J, Kansiime F (2006) Floating mats: their occurrence and influence on shoreline distribution of emergent vegetation. Freshwater Biology 51, 1286–1297

Azza NGT, Kansiime F, Nalubega M, Denny P (2000) Differential permeability of papyrus and Miscanthidium root mats in Nakivubo swamp, Uganda. Aquat Bot 67:167–178

Bachelet D, Hunt HW, Detling JK (1989) A simulation model of intraseasonal carbon and nitrogen dynamics of blue grama swards as influenced by above- and belowground grazing. Ecol. Modelling 44: 231-252

Balasubramanian M (2019) Economic value of regulating ecosystem services: a comprehensive at the global level review. Environmental Monitoring and Assessment 191(10):616

Becht R, Odada EO, Higgins S (2006) Lake Naivasha - Experience and lessons learnt brief. In: Managing lakes and their basins for sustainable use. International Lake Environment Committee Foundation, Kosatsu, Japan (available at: http://www.ilec.or.jp/eg/pubs/index.html)

Benjankar R, Egger G, Jorde K, Goodwin P, Glenn NF (2011) Dynamic floodplain vegetation model development for the Kootenai River, USA. Journal of Environmental Management 92(12):3058-3070

Benjankar R, Glenn NF, Egger G, Jorde K, Goodwin P (2010) Comparison of field-observed and simulated map output from a dynamic floodplain vegetation model using remote sensing and GIS techniques. GIScience & Remote Sensing 47(4):480-497

Beuel S, Alvarez M, Amler E, Behn K, Kotze D, Kreye C, Leemhuis C, Wagner K, Willy DK, Ziegler S, Becker M (2016) A rapid assessment of anthropogenic disturbances in East African wetlands. Ecological Indicators 67:684-692

Beusen AHW, Bouwman AF, Van Beek LPJH, Mogollón JM, Middelburg JJ (2016) Global riverine N and P transport to ocean increased during the 20th century despite increased retention along the aquatic continuum. Biogeosciences 13, 2441–2451

Boar RR (2006) Respones of a fringing *Cyperus papyrus* L. swamp to changes in water level. Aquat Bot 84: 85-92

Boar RR, Harper DM (2002) Magnetic susceptibilities of lake sediment and soils on the shoreline of Lake Naivasha, Kenya. Hydrobiologia 488: 81-88

Boar RR, Harper DM, Adams CS (1999) Biomass allocation in *Cyperus papyrus* in a tropical Wetland, Lake Naivasha, Kenya. Biotropica 31(3): 411-421

Bonan GB, Doney SC (2018) Climate, ecosystems, and planetary futures: The challenge to predict life in Earth system models. Science 359(6375):eaam8328

Bousquet P, Ciais P, Miller JB, Dlugokencky EJ, Hauglustaine DA, Prigent C, Van der Werf GR, Peylin P, Brunke EG, Carouge C, Langenfelds RL, Lathière J, Papa F, Ramonet M, Schmidt M, Steele LP, Tyler SC, White J (2006) Contribution of anthropogenic and natural sources to atmospheric methane variability. Nature 443(7110):439-443

Bradshaw CJA, Ehrlich PR, Beattie A, Ceballos G, Crist E, Diamond J, Dirzo R, Ehrlich AH, Harte J, Harte ME, Pyke G, Raven PH, Ripple WJ, Saltré F, Turnbull C, Wackernagel M, Blumstein DT (2021) Underestimating the Challenges of Avoiding a Ghastly Future. Frontiers in Conservation Science 1(9). DOI:10.3389/fcosc.2020.615419

Brander L, Brouwer R, Wagtendonk A (2013) Economic valuation of regulating services provided by wetlands in agricultural landscapes: A meta-analysis. Ecological Engineering 56:89-96

Brauman KA, Daily GC, Duarte TK, Mooney HA (2007) The Nature and Value of Ecosystem Services: An Overview Highlighting Hydrologic Services. Annu Rev Env Resour 32(1):67-98

Bravo HR, Jiang F, Hunt RJ (2002) Using groundwater temperature data to constrain parameter estimation in a groundwater flow model of a wetland system. Water Resources Research 38(8):28-1-28-14

Bregoli F, Crosato A, Paron P, McClain ME (2019) Humans reshape wetlands: Unveiling the last 100 years of morphological changes of the Mara Wetland, Tanzania. Science of the Total Environment 691: 896–907

Bridgham SD, Moore TR, Richardson CJ, Roulet NT (2014) Errors in greenhouse forcing and soil carbon sequestration estimates in freshwater wetlands: a comment on Mitsch et al. (2013). Landscape ecology 29(9):1481-1485

Brooks RP, Wardrop DH, Cole CA, Campbell DA (2005) Are we purveyors of wetland homogeneity?: A model of degradation and restoration to improve wetland mitigation performance. Ecological Engineering 24(4):331-340

Butchart SHM, Walpole M, Collen B, van Strien A, Scharlemann JPW, Almond EA, Baillie JEM, Bomhard B, Brown C, Bruno J, Carpenter KE, Carr GM, Chanson J, Chenery AM, Csirke J, Davidson NC, Dentener F, Foster M, Galli A, Galloway JN, Genovesi P, Gregory RD, Hockings M, Kapos V, Lamarque J-F, Leverington F, Loh J, McGeoch MA, McRae L, Minasyan A, Morcillo MH, Oldfield TEE, Pauly D, Quader S, Revenga C, Sauer JR, Skolnik B, Spear D, Stanwell-Smith D, Stuart SN, Symes A, Tierney M, Tyrrell TD, Vié J-C, Watson R (2010) Global Biodiversity: Indicators of Recent Declines. Science 328: 1164–1168

Carballeira R, Souto M (2018) Presencia de *Cyperus papyrus* L. (Cyperaceae) en la región biogeográfica atlántica de la Península Ibérica. Acta Botanica Malacitana 43:137-140

Cariboni J, Gatelli D, Liska R, Saltelli A (2007) The role of sensitivity analysis in ecological modelling. Ecol Model 203:167-182

Carpenter SR, Mooney HA, Agard J, Capistrano D, DeFries RS, Díaz S, Dietz T, Duraiappah AK, Oteng-Yeboah A, Pereira HM, Perrings C, Reid WV, Sarukhan J, Scholes RJ, Whyte A (2009) Science for managing ecosystem services: beyond the Millennium Ecosystem Assessment. PNAS 106: 1305-1312

Ch. Munch P (1861) Bemerkungen über den *Cyperus Papyrus* Lin. Oesterreichische Botanische Zeitschrift 11(11):364-367

Chale FMM (1987) Plant biomass and nutrient levels of a tropical macrophyte (*Cyperus papyrus* L.) receiving domestic wastewater. Hydrobiological Bulletin 21: 167-170

Chavan PV, Dennett KE (2008) Wetland Simulation Model for Nitrogen, Phosphorus, and Sediments Retention in Constructed Wetlands. Water, Air, and Soil Pollution 187(1):109-118

Chen R, Twilley RR (1999) A simulation model of organic matter and nutrient accumulation in mangrove wetland soils. Biogeochemistry 44(1):93-118

Chimney MJ, Pietro KC (2006) Decomposition of macrophyte litter in a subtropical constructed wetland in south Florida (USA). Ecological Engineering 27:301-321

Chow-Fraser P (1998) A conceptual ecological model to aid restoration of Cootes Paradise Marsh, a degraded coastal wetland of Lake Ontario, Canada. Wetlands Ecology and Management 6(1):43-57

Conceição P, Levine S, Lipton M, Warren-Rodríguez A (2016) Toward a food secure future: Ensuring food security for sustainable human development in Sub-Saharan Africa. Food Pol 60:1-9

Costanza R, de Groot R, Braat L, Kubiszewski I, Fioramonti L, Sutton P, Farber S, Grasso M (2017) Twenty years of ecosystem services: How far have we come and how far do we still need to go. Ecosyst Serv 28:1-16

Costanza R, Sklar FH, White ML (1990) Modeling Coastal Landscape Dynamics. BioScience 40(2):91-107

Costanza R, Voinov A eds. (2004) Landscape simulation modeling: a spatially explicit, dynamic approach. Springer-Verlag, New York, USACózar A, Bergamino N, Mazzuoli S, Azza N, Bracchini L, Dattilo AM, Loiselle SA (2007) Relationships between wetland ecotones and inshore water quality in the Ugandan coast of Lake Victoria Wetlands Ecol Manage 15: 499-507. DOI 10.1007/s11273-007-9046-6

Chimney MJ, Pietro KC (2006) Decomposition of macrophyte litter in a subtropical constructed wetland in south Florida (USA). Ecological Engineering 27: 301-321

Clevering OA, Brix H, Lukavská J (2001) Geographic variation in growth responses in *Phragmites australis*. Aquatic Botany 69: 89-108

Cui J, Li C, Trettin C (2005) Analyzing the ecosystem carbon and hydrologic characteristics of forested wetland using a biogeochemical process model. Global Change Biology 11(2):278-289

Darwall WRT, Bremerich V, De Wever A, Dell AI, Freyhof J, Gessner MO, Grossart H-P, Harrison I, Irvine K, Jähnig SC, Jeschke JM, Lee JJ, Lu C, Lewandowska AM, Monaghan MT, Nejstgaard JC, Patricio H, Schmidt-Kloiber A, Stuart SN, Thieme M, Tockner K, Turak E, Weyl O (2018) The Alliance for Freshwater Life: A global call to unite efforts for freshwater biodiversity science and conservation. Aquatic Conservation 28:1015-1022

Davidson NC (2014) How much wetland has the world lost? Long-term and recent trends in global wetland area. Mar Freshwater Res 65(10):934-941

Davidson NC, Dinesen L, Fennessy S, Finlayson CM, Grillas P, Grobicki A, McInnes RJ, Stroud DA (2020) Trends in the ecological character of the world's wetlands. Mar. Freshw. Res. 71, 127

Davidson NC, Finlayson CM (2018) Extent, regional distribution and changes in area of different classes of wetland. Mar Freshw Res 69:1525-1533

Davidson NC, Fluet-Chouinard E, Finlayson CM (2018) Global extent and distribution of wetlands: trends and issues. Marine and Freshwater Research 69(4):620-627

Davidson NC, van Dam AA, Finlayson CM, Mcinnes RJ (2019) Worth of wetlands: Revised global monetary values of coastal and inland wetland ecosystem services. Mar. Freshw. Res. 70, 1189–1194

Day JW, Rybczyk J, Scarton F, Rismondo A, Are D, Cecconi G (1999) Soil Accretionary Dynamics, Sea-Level Rise and the Survival of Wetlands in Venice Lagoon: A Field and Modelling Approach. Estuarine, Coastal and Shelf Science 49(1):607-628

De Groot RS, Brander L, Finlayson CM (2018) Wetland Ecosystem Services. In: The Wetland Book I: Structure and Function, Management and Methods. Finlayson CM, Everard M, Irvine K, McInnes RJ, Middleton BA, van Dam AA, Davidson NC (eds). Springer Netherlands, Dordrecht: 323-333

De Groot RS, Brander L, van der Ploeg S, Costanza R, Bernard F, Braat L, Christie M, Crossman N, Ghermandi A, Hein L, Hussain S, Kumar P, McVittie A, Portela R, Rodriguez LC, ten

Brink P, van Beukering P (2012) Global estimates of the value of ecosystems and their services in monetary units. Ecosystem Services 1(1):50-61

De Groot RS, Fisher B, Christie M, Aronson J, Braat LR, Haines-Young, Gowdy J, Maltby E, Neuville A. Polasky S, Portela R, Ring I (2010) Integrating the ecological and economic dimensions in biodiversity and ecosystem service valuation. In: Kumar P (Ed.) TEEB Foundations, The Economics of Ecosystems and Biodiversity: Ecological and Economic Foundations (Chapter 1). Earthscan, London

De Groot RS, Stuip MAM, Finlayson CM, Davidson N (2006) Valuing wetlands: guidance for valuing the benefits derived from wetland ecosystem services, Ramsar Technical Report No. 3/CBD Technical Series No. 27. Ramsar Convention Secretariat, Gland, Switzerland & Secretariat of the Convention on Biological Diversity, Montreal, Canada

DeAngelis DL, Mooij WM (2005) Individual-based modeling of ecological and evolutionary processes. Annu. Rev. Ecol. Evol. Syst. 36:147-168

Deegan BM, White SD, Ganf GG (2007) The influence of water level fluctuations on the growth of four emergent macrophyte species. Aquatic Botany 86: 309-315

Denny P (1984) Permanent swamp vegetation of the Upper Nile. Hydrobiologia 110:79–90

Denny P (ed) (1985) The ecology and management of African wetland vegetation. Dr W Junk Publishers, Dordrecht

Dentener F, Drevet J, Lamarque JF, Bey I, Eickhout B, Fiore AM, Hauglustaine D, Horowitz LW, Krol M, Kulshrestha UC, Lawrence M, Galy-Lacaux C, Rast S, Shindell D, Stevenson D, Van Noije T, Atherton C, Bell N, Bergman D, Butler T, Cofala J, Collins B, Doherty R, Ellingsen K, Galloway J, Gauss M, Montanaro V, Müller JF, Pitari G, Rodriguez J, Sanderson M, Solmon F, Strahan S, Schultz M, Sudo K, Szopa S, Wild O (2006) Nitrogen and sulfur deposition on regional and global scales: A multimodel evaluation. Global Biogeochemical Cycles 20, 1-21. GB4003, doi: 10.1029/ 2005GB002672

Díaz S, Demissew S, Carabias J, Joly C, Lonsdale M, Ash N, Larigauderie A, Adhikari JR, Arico S, Báldi A, Bartuska A, Baste IA, Bilgin A, Brondizio E, Chan KMA, Figueroa VE, Duraiappah A, Fischer M .... Zlatanova D (2015) The IPBES Conceptual Framework — connecting nature and people. Current Opinion in Environmental Sustainability 14:1-16

Dixon MJR, Loh J, Davidson NC, Beltrame C, Freeman R, Walpole M (2016) Tracking global change in ecosystem area: The Wetland Extent Trends index. Biol Conserv:193:27-35

Driver PD, Barbour EJ, Michener K (2011) An integrated surface water, groundwater and wetland plant model of drought response and recovery for environmental water management. 19th International Congress on Modelling and Simulation, Perth, Australia: pp12-16

Elser JJ, Andersen T, Baron JS, Bergström AK, Jansson M, Kyle M, Nydick KR, Steger L, Hessen DO (2009) Shifts in Lake N:P Stoichiometry and Nutrient Limitation Driven by Atmospheric Nitrogen Deposition. Science 5954(326):835-837

Elser JJ, Fagan WF, Kerkhoff AJ, Swenson NG, Enquist BJ (2010) Biological stoichiometry of plant production: metabolism, scaling and ecological response to global change New Phytologist 186:593-608

Emerton L (ed) (2005) Values and rewards: counting and capturing ecosystem water services for sustainable development. IUCN Water, Nature and Economics Technical Paper No. 1. The World Conservation Union, Ecosystems and Livelihoods Group Asia

EOS 2017 DNDC (Version 9.5) Scientific Basis and Processes. Institute for the Study of Earth, Oceans, and Space University of New Hampshire, Durham, USA

Erb K-H, Luyssaert S, Meyfroidt P, Pongratz J, Don A, Kloster S, Kuemmerle T, Fetzel T, Fuchs R, Herold M, Haberl H, Jones CD, Marín-Spiotta E, McCallum I, Robertson E, Seufert V, Fritz S, Valade A, Wiltshire A, Dolman AJ (2017) Land management: data availability and process understanding for global change studies. Global Change Biology 23(2):512-533

Evrard O, Nord G, Cerdan O, Souchère V, Le Bissonnais Y, Bonté P (2010) Modelling the impact of land use change and rainfall seasonality on sediment export from an agricultural catchment of the northwestern European loess belt. Agriculture, Ecosystems & Environment 138(1):83-94

Fan Y, Miguez-Macho G (2011) A simple hydrologic framework for simulating wetlands in climate and earth system models. Climate Dynamics 37(1):253-278

FAO (2019) The State of the World's Biodiversity for Food and Agriculture. Bélanger J, Pilling D (eds), FAO Commission on Genetic Resources for Food and Agriculture Assessments, Rome

FAO-IWMI (2018) More people, more food, worse water? A global review of water pollution from agriculture. Mateo-Sagasta J, Zadeh SM, Turral H (eds), the Food and Agriculture Organisation of the United Nations, Rome, Italy and the International Water Management Institute, Colombo, Sri Lanka

Feng M, Liu S, Euliss NH, Young C, Mushet DM (2011) Prototyping an online wetland ecosystem services model using open model sharing standards. Environmental Modelling and Software 26(4):458-468

Feng K, Molz FJ (1997) A 2-D, diffusion-based, wetland flow model. Journal of Hydrology 196(1):230-250

Finlayson CM (2012) Forty years of wetland conservation and wise use. Aquat Conserv: Mar Freshw Ecosyst 22:139-143

Finlayson CM (2018) Climate Change and Wetlands. In: The Wetland Book I: Structure and Function, Management and Methods. Finlayson CM, Everard M, Irvine K, McInnes RJ, Middleton BA, van Dam AA, Davidson NC (eds). Springer Netherlands, Dordrecht: 597-608

Finlayson CM, Davies GT, Moomaw WR, Chmura GL, Natali SM, Perry JE, Roulet N, Sutton-Grier AE (2019) The Second Warning to Humanity – Providing a Context for Wetland Management and Policy. Wetlands 39(1):1-5

Fisher J, Acreman MC (2004) Wetland nutrient removal: a review of the evidence. Hydrology and EarthSystem Sciences Discussions, European Geosciences Union, 8(4):673-685

Fitz HC, DeBellevue EB, Costanza R, Boumans R, Maxwell T, Wainger L, Sklar FH (1996) Development of a general ecosystem model for a range of scales and ecosystems. Ecological Modelling 88(1):263-295

Fitz HC, Hughes N (2008) Wetland Ecological Models. EDIS 2008(3)

Gabrielsson S, Brogaard S, Jerneck A (2013) Living without buffers—illustrating climate vulnerability in the Lake Victoria basin. Sustainability Science 8: 143-157

Galloway JN, Dentener FJ, Capone DG, Boyer EW, Howarth RW, Seitzinger SP, Asner GP, Cleveland CC, Green PA, Holland EA, Karl DM, Michaels AF, Porter JH, Townsend AR,

Vorosmarty CJ (2004) Nitrogen cycles: past, present, and future. Biogeochemistry 70: 153-226

Gallardo M, Elia A, Thompson RB (2020) Decision support systems and models for aiding irrigation and nutrient management of vegetable crops. Agricultural Water Management 240:106209

Gaudet JJ (1975) Mineral concentrations in papyrus in various African swamps. J Ecol 63(2):483–491

Gaudet JJ (1977a) Natural drawdown on Lake Naivasha, Kenya, and the formation of papyrus swamps. Aquat Bot 3(C):1-47

Gaudet JJ (1977b) Uptake, accumulation, and loss of nutrients by papyrus in tropical swamps. Ecology 58, 415-422

Gaudet JJ (1979) Seasonal changes in nutrients in a tropical swamp: North Swamp, Lake Naivasha, Kenya. J Ecol 67:953–981

Gaudet JJ, Muthuri FM (1981) Nutrient relationships in shallow water in an African Lake, Lake Naivasha. Oecologia 49(1):109-118

Geheb K, Binns T (1997) 'Fishing farmers' of 'farming fishermen'? The quest for household income and nutritional security on the Kenyan shores of Lake Victoria. African Affairs 96, 73-93

Geremew A, Stiers I, Sierens T, Kefalew A, Triest L (2018) Clonal growth strategy, diversity and structure: A spatiotemporal response to sedimentation in tropical *Cyperus papyrus* swamps. PLoS ONE 13: e0190810

Gettel GM, Namaalwa S, van Dam AA (2019) Agricultural conversion of papyrus wetlands leads to loss of denitrification potential, Namatala wetland, Uganda. Presented at the 8th International Symposium on Wetland Pollutant Dynamics and Control - WETPOL2019, 17-21 June, 2019, Aarhus, Denmark

Gettel GM, Tshering K, Nakitende H, van Dam AA (2012) Tradeoffs in regulating ecosystem services in East African Papyrus Wetlands: Denitrification as a case study. In AGU Fall Meeting Abstracts 1:6

Gichuki CM, Gichuki NN (1992) Wetland birds of Kenya. p. 37-46, In: Crafter, S.A., Njuguna, S.G. and Howard, G.W. (eds.) Wetlands of Kenya. Proceedings of the KWWG Seminar on Wetlands of Kenya, National Museums of Kenya, Nairobi, 3-5 July 1991. IUCN, Nairobi

Gichuki J, Guebas FD, Mugo J, Rabuor CO, Triest L, Dehairs F (2001) Species inventory and the local uses of the plants and fishes of the Lower Sondu Miriu wetland of Lake Victoria, Kenya. Hydrobiologia 458(1):99-106

Gichuki JW, Triest L, Dehairs F (2005) The fate of organic matter in a papyrus (*Cyperus papyrus* L.) dominated tropical wetland ecosystem in Nyanza Gulf (Lake Victoria, Kenya) inferred from δ13C and δ15N analysis. Isotopes in Environmental and Health Studies 41: 379 - 390. DOI: 10.1080/10256010500384739

Giltrap DL, Li C, Saggar S (2010) DNDC: A process-based model of greenhouse gas fluxes from agricultural soils. Agriculture, Ecosystems & Environment 136(3):292-300

Gordon LJ, Finlayson CM, Falkenmark M (2010) Managing water in agriculture for food production and other ecosystem services. Agricultural Water Management 97(4):512-519

Grace JB, Pugesek BH (1997) A structural equation model of plant species richness and its application to a coastal wetland. The American Naturalist 149(3):436-460

Grime JP (1974) Vegetation classification by reference to strategies. Nature 250: 26-31

Grime JP (1977) Evidence for the Existence of Three Primary Strategies in Plants and Its Relevance to Ecological and Evolutionary Theory. The American Naturalist 111: 1169-1194

Guerry AD, Polasky S, Lubchenco J, Chaplin-Kramer R, Daily GC, Griffin R, Ruckelshaus M, Bateman IJ, Duraiappah A, Elmqvist T, Feldman MW (2015) Natural capital and ecosystem services informing decisions: From promise to practice. P Natl Acad Sci USA 112(24):7348-7355

Guildford SJ, Hecky RE (2000) Total nitrogen, total phosphorus, and nutrient limitation in lakes and oceans: Is there a common relationship? Limnol Oceanogr 45(6):1213-1223

Haberl H, Winiwarter V, Andersson K, Ayres RU, Boone C, Castillo A, Cunfer G, Fischer-Kowalski M, Freudenburg WR, Furman E, Kaufmann R, Krausmann F, Langthaler E, Lotze-Campen H, Mirtl M, Redman CL, Reenberg A, Wardell A, Warr B, Zechmeister H (2006) From LTER to LTSER: Conceptualizing the Socioeconomic Dimension of Long-term Socioecological Research. Ecol Soc 11(2):13

Haddis A, van der Bruggen B, Smets I (2020) Constructed wetlands as nature based solutions in removing organic pollutants from wastewater under irregular flow conditions in a tropical climate. Ecohydrology & Hydrobiology 20(1):38-47

Hammer DE, Kadlec RH (1986) A model for wetland surface water dynamics. Water Resources Research 22(13):1951-1958

Hecky RE, Mugidde R, Ramlal PS, Talbot MR, Kling GW (2010) Multiple stressors cause rapid ecosystem change in Lake Victoria. Freshwater Biology 55: 19-42

Hickley P, Muchiri M, Boar R, Britton R, Adams C, Gichuru N, Harper D (2004) Habitat degradation and subsequent fishery collapse in Lakes Naivasha and Baringo, Kenya. Ecohydrology and Hydrobiology 4: 503-517

Hiraishi T, Krug T, Tanabe K, Srivastava N, Baasansuren J, Fukuda M, Troxler T (eds) (2013) Supplement to the 2006 IPCC guidelines for national greenhouse gas inventories: wetlands. IPCC, Geneva

Hudon C, Wilcox D, Ingram J (2006) Modeling wetland plant community response to assess water-level regulation scenarios in the Lake Ontario–St. Lawrence River basin. Environmental Monitoring and Assessment 113(1-3):303-328

Hughes DA, Tshimanga RM, Tirivarombo S, Tanner J (2014) Simulating wetland impacts on stream flow in southern Africa using a monthly hydrological model. Hydrological Processes 28(4):1775-1786

IPBES (2019) Global assessment report on biodiversity and ecosystem services of the Intergovernmental Science-Policy Platform on Biodiversity and Ecosystem Services. Brondizio ES, Settele J, Díaz S, Ngo HT (eds). IPBES secretariat, Bonn

IPCC (2014) Climate Change 2014: Synthesis Report. Contribution of Working Groups I, II and III to the Fifth Assessment Report of the Intergovernmental Panel on Climate Change. Pachauri RK, Meyer LA (eds). IPCC, Geneva

Irvine K, Etiegni CA, Weyl OLF (2018) Prognosis for long- term sustainable fisheries in the African Great Lakes. Fish. Manag. Ecol. 1–13

Janse JH, Kuiper JJ, Weijters MJ, Westerbeek EP, Jeuken MHJL, Bakkenes M, Alkemade R, Mooij WM, Verhoeven JTA (2015) GLOBIO-Aquatic, a global model of human impact on the biodiversity of inland aquatic ecosystems. Environmental Science & Policy 48, 99-114.

Janse JH, Ligtvoet W, Van Tol S, Bresser AH (2001) A model study on the role of wetland zones in lake eutrophication and restoration. TheScientificWorldJournal 1 Suppl 2:605-614

Janse JH, van Dam AA, Hes EMA, de Klein JJM, Finlayson CM, Janssen ABG, van Wijk D, Mooij WM, Verhoeven JTA (2019) Towards a global model for wetlands ecosystem services. Curr Opin Env Sust 36:11-19

Janssen ABG, Arhonditsis GB, Beusen A, Bolding K, Bruce L, Bruggeman J, Couture RM, Downing AS, Elliott JA, Frassl MA, Gal G, Gerla DJ, Hipsey MR, Hu F, Ives SC, Janse JH, Jeppesen E, Johnk KD, Kneis D, Kong X, Kuiper JJ, Lehmann MK, Lemmen C, Ozkundakci D, Petzoldt T, Rinke K, Robson BJ, Sachse R, Schep SA, Schmid M, Scholten H, Teurlincx S, Trolle D, Troost TA, van Dam AA, van Gerven LPA, Weijerman M, Wells SA, Mooij WM (2015) Exploring, exploiting and evolving diversity of aquatic ecosystem models: a community perspective. Aquat Ecol 49:513–548

Jayne TS, Snapp S, Place F, Sitko N (2019) Sustainable agricultural intensification in an era of rural transformation in Africa. Global Food Security 20: 105-113

Ji ZG, Morton MR, Hamrick JM (2001) Wetting and Drying Simulation of Estuarine Processes. Estuarine, Coastal and Shelf Science 53(5):683-700

Johnson WC, Millett BV, Gilmanov T, Voldseth RA, Guntenspergen GR, Naugle DE (2005) Vulnerability of Northern Prairie Wetlands to Climate Change. BioScience 55(10)863-872

Johnston CA (1991) Sediment and nutrient retention by freshwater wetlands: Effects on surface water quality. Critical Reviews in Environmental Control 21: 491-565

Jones MB (1988) Photosynthetic responses of C3 and C4 wetland species in a tropical swamp. Journal of Ecology 76: 253-262

Jones MB, Humphries SW (2002) Impact of the C4 sedge *Cyperus papyrus* L. on carbon and water fluxes in an African wetland. Hydrobiologia 488: 107-133

Jones MB, Kansiime F, Saunders MJ (2016) The potential use of papyrus (*Cyperus papyrus* L.) wetlands as a source of biomass energy for sub-Saharan Africa. GCB Bioenergy doi: 10.1111/gcbb.12392

Jones MB, Muthuri FM (1985) The canopy structure and microclimate of papyrus (*Cyperus papyrus*) swamps. J Ecol 73(2):481-491

Jones MB, Muthuri FM (1997) Standing biomass and carbon distribution in a papyrus (*Cyperus papyrus* L.) swamp on Lake Naivasha, Kenya. J Trop Ecol 13: 347-356

Kabumbuli R, Kiwazi FW (2009) Participatory planning, management and alternative livelihoods for poor wetland-dependent communities in Kampala, Uganda. Afr J Ecol 47 (Suppl. 1): 154–160

Kadlec RH (1997) An autobiotic wetland phosphorus model. Ecological Modelling 8(2): 145-172

Kaggwa RC, Mulalelo CI, Denny P, Okurut TO (2001) The impact of alum discharges on a natural tropical wetland in Uganda. Water Res 35(3):795-807

Kansiime F, Kateyo E, Oryem-Origa H, Mucunguzi P (2007) Nutrient status and retention in pristine and disturbed wetlands in Uganda: Management implications. Wetl Ecol Manag 15(6):453-467

Kansiime F, Nalubega M (1999) Wastewater treatment by a natural wetland: the Nakivubo Swamp, Uganda. Process and Implications. PhD thesis, Agricultural University of Wageningen, Wageningen

Kansiime F, Nalubega M, van Bruggen JJ, Denny P (2003) The effect of wastewater discharge on biomass production and nutrient content of *Cyperus papyrus* and *Miscanthidium violaceum* in the Nakivubo wetland, Kampala, Uganda. Water Science and Technology 48: 233-40

Kanyiginya V, Kansiime F, Kimwaga R, Mashauri DA (2010) Assessment of nutrient retention by Natete wetland Kampala, Uganda. Phys Chem Earth 35(13-14):657-664

Kazezyılmaz-Alhan CM, Medina Jr MA, Richardson CJ (2007) A wetland hydrology and water quality model incorporating surface water/groundwater interactions. Water Resources Research 43(4): W04434

Kelderman P, Kansiime F, Tola MA, van Dam AA (2007) The role of sediments for phosphorous retention in the Kirinya wetland (Uganda). Wetl Ecol Manage 15: 481-488

Kengne IM, Akoa A, Soh EK, Tsama V, Ngoutane MM, Dodane PH, Koné D (2008) Effects of faecal sludge application on growth characteristics and chemical composition of *Echinochloa pyramidalis* (Lam.) Hitch. and Chase and *Cyperus papyrus* L. Ecological Engineering 34: 233-242

Khisa PS, Uhlenbrook S, van Dam AA, Wenninger J, Griensven A, Abira M (2013) Ecohydrological characterization of the Nyando wetland, Lake Victoria, Kenya: A State of System (SoS) analysis. African Journal of Environmental Science and Technology 7: 417-434

Kibwage JK, Onyango PO, Bakamwesiga H (2008) Local institutions for sustaining wetland resources and community livelihoods in the Lake Victoria basin. African Journal of Environmental Science and Technology 2(5):97-106

Kipkemboi J, Kansiime F, Denny P (2002) The response of *Cyperus papyrus* (L.) and *Miscanthidium violaceum* (K. Schum.) Robyns to eutrophication in natural wetlands of Lake Victoria, Uganda. Afr J Aquat Sci 27(1):11-20

Kipkemboi J, van Dam AA (2018) Papyrus Wetlands. In: The Wetland Book II: Distribution, Description, and Conservation. Finlayson CM, Milton GR, Prentice RC, Davidson NC (eds). Springer Netherlands, Dordrecht: 183-197

Kipkemboi J, van Dam AA, Ikiara MM, Denny P (2007) Integration of smallholder wetland aquaculture-agriculture systems (Fingerponds) into riparian farming systems at the shores of Lake Victoria, Kenya: socio-economics and livelihoods. Geogr J 173: 257–272

Kiwango Y, Moshi G, Kibasa W, Mnaya B (2013) Papyrus wetlands creation, a solution to improve food security and save Lake Victoria. Wetl Ecol Manag 21(2):147-154

Kiwango YA, Wolanski E (2008) Papyrus wetlands, nutrients balance, fisheries collapse, food security, and Lake Victoria level decline in 2000–2006. Wetland Ecology and Management 16: 89-96

Klimeš L (2000) *Phragmites australis* at an extreme altitude: rhizome architecture and its modelling. Folia Geobotanica 35(4):403-417

KNBS (2010) The 2009 Kenya Population and Housing Census. Kenya National Bureau of Statistics, Nairobi

Kollas C, Kersebaum KC, Nendel C, Manevski K, Müller C, Palosuo T, Armas-Herrera CM, Beaudoin N, Bindi M, Charfeddine M, Conradt T, Constantin J, Eitzinger J, Ewert F, Ferrise R, Gaiser T, Cortazar-Atauri I, Giglio L, Hlavinka P, Hoffmann H, Hoffmann MP, Launay M, Manderscheid R, Mary B, Mirschel W, Moriondo M, Olesen JE, Öztürk I, Pacholski A, Ripoche-Wachter D, Roggero PP, Roncossek S, Rötter RP, Ruget F, Sharif B, Trnka M, Ventrella D, Waha K, Wegehenkel M, Weigel H-J, Wu L (2015) Crop rotation modelling—A European model intercomparison. European Journal of Agronomy 70:98-111

Kotze D, Marneweck G, Batchelor A, Lindley D, Collins N (2009) WET-EcoServices: a technique for rapidly assessing ecosystem services supplied by wetlands. South African National Biodiversity Institute http://hdl.handle.net/20.500.12143/770

Kremen C (2015) Reframing the land-sparing/land-sharing debate for biodiversity conservation. Annals of the New York Academy of Sciences 1355: 52-76

Lagerwall G, Kiker G, Muñoz-Carpena R, Convertino M, James A, Wang N (2012) A spatially distributed, deterministic approach to modeling *Typha domingensis* (cattail) in an Everglades wetland. Ecological Processes 1: 10

Langan C, Farmer J, Rivington M, Smith JU (2018) Tropical wetland ecosystem service assessments in East Africa; A review of approaches and challenges. Environmental Modelling & Software 102:260-273

Laidig KJ, Zampella RA, Brown AM, Procopio NA (2010) Development of vegetation models to predict the potential effect of groundwater withdrawals on forested wetlands. Wetlands 30(3):489-500

Lamers M, Ingwersen J, Streck T (2007) Modelling nitrous oxide emission from water-logged soils of a spruce forest ecosystem using the biogeochemical model Wetland-DNDC. Biogeochemistry 86(3):287-299

Li C, Cui J, Sun G, Trettin C (2004) Modeling Impacts of Management on Carbon Sequestration and Trace Gas Emissions in Forested Wetland Ecosystems. Environmental Management 33(1):S176-S186

Li T, Hasegawa T, Yin X, Zhu Y, Boote K, Adam M, Bregaglio S, Buis S, Confalonieri R, Fumoto T, Gaydon D, Marcaida III M, Nakagawa H, Oriol P, Ruane AC, Ruget F, Singh B, Singh U, Tang L, Tao F, Wilkens P, Yoshida H, Zhang Z, Bouman B (2015) Uncertainties in predicting rice yield by current crop models under a wide range of climatic conditions. Global Change Biology 21(3):1328-1341

Lorenzen B, Brix H, Mendelssohn IA, McKee KL, Miao SL (2001) Growth, biomass allocation and nutrient use efficiency in *Cladium jamaicense* and *Typha domingensis* as affected phosphorus and oxygen availability. Aquatic Botany 70: 117-133

Loiselle SA, Azza N, Cózar A, Bracchini L, Tognazzi A, Dattilo A, Rossi C (2008) Variabillity in factors causing light attenuation in Lake Victoria. Freshw Biol 53: 535-545

Maclean IMD, Hassall M, Boar RR, Lake IR (2006) Effects of disturbance and habitat loss on papyrus-dwelling passerines. Biological Conservation 131, 349-358

Maclean IMD, Bird JP, Hassall M (2014) Papyrus swamp drainage and the conservation status of their avifauna. Wetl Ecol Manag 22:115-127

Makler-Pick V, Gal G, Gorfine M, Hipsey MR, Carmel Y (2011) Sensitivity analysis for complex ecological models – A new approach. Environ Modell Softw 26(2):124-134

Maltby E ed. (2009) Functional assessment of wetlands: towards evaluation of ecosystem services. Woodhead Publishing Limited, Oxford, Cambridge, New Delhi

Maltby E, Acreman MC (2011) Ecosystem services of wetlands: pathfinder for a new paradigm. Hydrol Sci J 56, 1341-1359

Mansell RS, Bloom SA, Sun G (2000) A model for wetland hydrology: description and validation. Soil Science 165(5):384-397

Manzoni S, Trofymow JA, Jackson RB, Porporato A (2010) Stoichiometric controls on carbon, nitrogen, and phosphorus dynamics in decomposing litter. Ecological Monographs 80: 89-106

Mburu N, Rousseau DP, van Bruggen JJ, Lens PN (2015) Use of the macrophyte *Cyperus papyrus* in wastewater treatment. In The Role of Natural and Constructed Wetlands in Nutrient Cycling and Retention on the Landscape (pp. 293-314). Springer, Cham

McInnes RJ (2018) Managing Wetlands for Water Supply. In: The Wetland Book I: Structure and Function, Management and Methods. Finlayson CM, Everard M, Irvine K, McInnes RJ, Middleton BA, van Dam AA, Davidson NC (eds). Springer Netherlands, Dordrecht: 1159-1165

MEA (2005) Ecosystems and Human Well-being: Wetlands and Water Synthesis. World Resources Institute, Washington DC

Melton JR, Wania R, Hodson EL, Poulter B, Ringeval B, Spahni R, Bohn T, Avis CA, Beerling DJ, Chen G, Eliseev AV (2013) Present state of global wetland extent and wetland methane modelling: conclusions from a model intercomparison project (WETCHIMP). Biogeosciences 10:753-788

Mitsch WJ, Bernal B, Nahlik AM, Mander Ü, Zhang L, Anderson CJ, Jørgensen SE, Brix H (2013) Wetlands, carbon, and climate change. Landscape Ecology 28(4):583-597

Mitsch WJ, Reeder BC (1991) Modelling nutrient retention of a freshwater coastal wetland: estimating the roles of primary productivity, sedimentation, resuspension and hydrology. Ecological Modelling 54(3):151-187

Mitsch WJ, Wang N (2000) Large-scale coastal wetland restoration on the Laurentian Great Lakes: Determining the potential for water quality improvement. Ecological Engineering 15(3):267-282

Mitchell SA (2013) The status of wetlands, threats and the predicted effect of global climate change: the situation in Sub-Saharan Africa. Aquatic sciences 75(1):95-112

Mnaya B, Asaeda T, Kiwango Y, Ayubu E (2007) Primary production in papyrus (*Cyperus papyrus* L.) of Rubondo Island, Lake Victoria, Tanzania. Wetl Ecol Manag 15(4):269-275

Mnaya B, Mtahiko MGG, Wolanski E (2017) The Serengeti will die if Kenya dams the Mara River. Oryx 51(4):581-583

Mooij WM, D Trolle, E Jeppesen, G Arhonditsis, PV Belolipetsky, DBR Chitamwebwa, AG Degermendzhy, DL DeAngelis, LN De Senerpont Domis, AS Downing, JA Elliott, CR Fragoso Jr, U Gaedke, SN Genova, RD Gulati, L Håkanson, DP Hamilton, MR Hipsey, J 't Hoen, S Hülsmann, FJ Los, V Makler, T Petzoldt, IG Prokopkin, K Rinke, SA Schep, K Tominaga, AA van Dam, EH van Nes, SA Wells and JH Janse (2010) Challenges and opportunities for integrating lake ecosystem modelling approaches. Aquatic Ecology 44:633-667

Mooij WM, van Wijk D, Beusen AHW, Brederveld RJ, Chang M, Cobben MMP, DeAngelis DL, Downing AS, Green P, Gsell AS, Huttunen I, Janse JH, Janssen ABG, Hengeveld GM, Kong X, Kramer L, Kuiper JJ, Langan SJ, Nolet BA, Nuijten RJM, Strokal M, Troost TA, van Dam AA, Teurlincx S (2019) Modeling water quality in the Anthropocene: Directions for the next-generation aquatic ecosystem models. Current Opinion in Environmental Sustainability 36: 85-95

Moomaw WR, Chmura GL, Davies GT, Finlayson CM, Middleton BA, Natali SM, Perry JE, Roulet N, Sutton-Grier AE (2018) Wetlands In a Changing Climate: Science, Policy and Management. Wetlands 38(2):183-205

Morrison EHJ, Banzaert A, Upton C, Pacini N, Pokorný J, Harper DM (2014) Biomass briquettes: A novel incentive for managing papyrus wetlands sustainably? Wetl Ecol Manag 22(2):129-141

Morrison EHJ, Upton C, Odhiambo-K'oyooh K, Harper DM (2012) Managing the natural capital of papyrus within riparian zones of Lake Victoria, Kenya. Hydrobiologia 692(1):5-17

Moustafa MZ, Hamrick JM (2000) Calibration of the Wetland Hydrodynamic Model to the Everglades Nutrient Removal Project. Water Quality and Ecosystem Modeling 1(1):141-167

Mugisha P, Kansiime F, Mucunguzi P, Kateyo E (2007) Wetland vegetation and nutrient retention in Nakivubo and Kirinya wetlands in the Lake Victoria basin of Uganda. Phys Chem Earth 32(15-18):1359-1365

Muraza M, Mayo AW, Norbert J (2013) Wetland Plant Dominance, Density and Biomass in Mara River Basin Wetland Upstream of Lake Victoria in Tanzania. International Journal of Scientific & Technology Research 2: 348-359

Muthuri FM, Jones MB (1997) Nutrient distribution in a papyrus swamp. Lake Naivasha, Kenya. Aquatic Botany 56, 35-50

Muthuri FM, Jones MB, Imbamba SK (1989) Primary productivity of papyrus (*Cyperus papyrus*) in a tropical swamp; Lake Naivash, Kenya. Biomass 18, 1-14

Muthuri FM, Kinyamario JI (1989) Nutritive Value of Papyrus (*Cyperus papyrus*, *Cyperaceae*), a Tropical Emergent Macrophyte. Economic Botany 43: 23-30

Mwakubo SM, Obare GA (2009) Vulnerability, livelihood assets and institutional dynamics in the management of wetlands in Lake Victoria watershed basin. Wetlands Ecology and Management 17, 613-626

Mwanuzi F, Aalderink H, Mdamo L (2003) Simulation of pollution buffering capacity of wetlands fringing the Lake Victoria. Environment International 29, 95-103

Mwaura FB, Widdowson D (1992) Nitrogenase activity in the papyrus swamps of Lake Naivasha, Kenya. Hydrobiologia 232:23–30

Nakayama T (2016) New perspective for eco-hydrology model to constrain missing role of inland waters on boundless biogeochemical cycle in terrestrial–aquatic continuum. Ecohydrology & Hydrobiology 16(3):138-148

Nakayama T (2017) Scaled-dependence and seasonal variations of carbon cycle through development of an advanced eco-hydrologic and biogeochemical coupling model. Ecological Modelling 356:151-161

Namaalwa S, van Dam AA, Funk A, Ajie GS, Kaggwa RC (2013) A characterization of the drivers, pressures, ecosystem functions and services of Namatala wetland, Uganda. Environmental Science and Policy 34:44-57

Namaalwa S, van Dam AA, Gettel GM, Kaggwa R, Zsuffa I, Irvine K (2020) The impact of wastewater discharge and agriculture on water quality and nutrient retention of Namatala wetland, Eastern Uganda. Frontiers in Environmental Science 8: 10.3389/fenvs.2020.00148

Noon KF (1996) A model of created wetland primary succession. Landscape and Urban Planning 34(2):97-123

Nyenje PM, Foppen JW, Uhlenbrook S, Kulabako R, Muwanga A (2010) Eutrophication and nutrient release in urban areas of sub-Saharan Africa — A review. Sci Total Environ 408(3):447-455

OECD/FAO (2016) "Agriculture in Sub-Saharan Africa: Prospects and challenges for the next decade". In OECD-FAO Agricultural Outlook 2016-2025, OECD Publishing, Paris

Ojoyi MM (2006) Sustainable Use of *Cyperus papyrus* at Lake Victoria wetlands in Kenya: A case study of Dunga and Kusa swamps. Sustainability Institute: Tecnologico' de Monterrey, Mexico

Olokotum M, Mitroi V, Troussellier M, Semyalo R, Bernard C, Montuelle B, Okello W, Quiblier C, Humbert JF (2020) A review of the socioecological causes and consequences of cyanobacterial blooms in Lake Victoria. Harmful Algae 96:101829

Opio A, Jones MB, Kansiime F, Otiti T (2014) Growth and Development of *Cyperus papyrus* in a Tropical Wetland. Open Journal of Ecology 4:113-123

Opio A, Jones MB, Kansiime F, Otiti T (2017) Response of *Cyperus papyrus* productivity to changes in relative humidity, temperature and photosynthetically active radiation. African Journal of Plant Science 11: 133-141

Osorio AM (2018) Modelling biomass production and nutrient retention of emergent macrophytes in wetland ecosystems. MSc thesis UNESCO-IHE Institute for Water Education, Delft, the Netherlands

Osumba JJL, Okeyo-Owuor JB, Raburu PO (2010) Effect of harvesting on temporal papyrus (*Cyperus papyrus*) biomass regeneration potential among swamps in Winam Gulf wetlands of Lake Victoria Basin, Kenya. Wetl Ecol Manag 18(3):333-341

Owino AO, Ryan PG (2007) Recent papyrus swamp habitat loss and conservation implications in western Kenya. Wetlands Ecology and Management 15(1):1-12

Pacini N, Hesslerová P, Pokorný J, Mwinami T, Morrison EHJ, Cook AA, Zhang S, Harper DM (2018) Papyrus as an ecohydrological tool for restoring ecosystem services in Afrotropical wetlands. Ecohydrology & Hydrobiology 18: 142-154

Peel MC, Finlayson BL, McMahon TA (2007) Updated world map of the Köppen-Geiger climate classification. Hydrology and Earth System Sciences Discussions 4(2):439-473

Peh KSH, Balmford A, Bradbury RB, Brown C, Butchart SHM, Hughes FMR, Stattersfield A, Thomas DHL, Walpole M, Bayliss J, Gowing D, Jones JPG, Lewis SL, Mulligan M, Pandeya B, Stratford C, Thompson JR, Turner K, Vira B, Willcock S, Birch JC (2013) TESSA: A toolkit for rapid assessment of ecosystem services at sites of biodiversity conservation importance. Ecosystem Services 5:51-57

Pendleton L, Donato DC, Murray BC, Crooks S, Jenkins WA, Sifleet S, Craft C, Fourqurean JW, Kauffman JB, Marbà N (2012) Estimating global "blue carbon" emissions from conversion and degradation of vegetated coastal ecosystems. PloS one 7(9):e43542

Perbangkhem T, Polprasert C (2010) Biomass production of papyrus (*Cyperus papyrus*) in constructed wetland treating low-strength domestic wastewater. 101: 833-835

Pianka ER (1970) On r- and K-Selection. The American Naturalist 104: 592-597

Pina-Ochoa E, Álvarez-Cobelas M (2006) Denitrification in aquatic environments: a cross-system analysis. Biogeochemistry 81: 111-130

Pohl C, Hirsch Hadorn G (2008) Methodological challenges of transdisciplinary research. Natures Sciences Sociétés 16(2):111-121

Pongratz J, Dolman H, Don A, Erb K-H, Fuchs R, Herold M, Jones C, Kuemmerle T, Luyssaert S, Meyfroidt P, Naudts K (2018) Models meet data: Challenges and opportunities in implementing land management in Earth system models. Global Change Biology 24(4):1470-1487

Poulin B, Davranche A, Lefebvre G (2010) Ecological assessment of *Phragmites australis* wetlands using multi-season SPOT-5 scenes. Remote Sensing of Environment 114(7):1602-1609

Powers SM, Johnson RA, Stanley EH (2012) Nutrient Retention and the Problem of Hydrologic Disconnection in Streams and Wetlands. Ecosystems 15(3):435-449

Pretty J, Toulmin C, Williams S (2011) Sustainable intensification in African agriculture. Int J Agric Sustain 9(1):5-24

R Core Team (2018) R: A language and environment for statistical computing. R Foundation for Statistical Computing, Vienna, Austria. URL https://www.R-project.org/.

R Core Team (2020) R: A language and environment for statistical computing. R Foundation for Statistical Computing, Vienna, Austria. URL https://www.R-project.org/.

Ramsar Convention on Wetlands (2018a) Global Wetland Outlook: State of the World's Wetlands and their Services to People. Ramsar Convention Secretariat. Gland

Ramsar Convention on Wetlands (2018b) Scaling up wetland conservation, wise use and restoration to achieve the Sustainable Development Goals – Wetlands and the SDGs. Ramsar Convention Secretariat. Gland

Raudsepp-Hearne C, Peterson GD, Bennett EM (2010) Ecosystem service bundles for analyzing tradeoffs in diverse landscapes. P Natl Acad Sci USA 107(11):5242-5247

Rebelo LM, McCartney MP, Finlayson CM (2010) Wetlands of Sub-Saharan Africa: distribution and contribution of agriculture to livelihoods. Wetl Ecol Manag 18(5):557-572

Reid GK, Lefebvre S, Filgueira R, Robinson SMC, Broch OJ, Dumas A, Chopin TBR (2020) Performance measures and models for open-water integrated multi-trophic aquaculture. Reviews in Aquaculture 12(1):47-75

Rejmánková E (2005) Nutrient resorption in wetland macrophytes: comparison across several regions of different nutrient status. New Phytol 167(2):471-482

Rejmánková E, Snyder JM (2008) Emergent macrophytes in phosphorus limited marshes: do phosphorus usage strategies change after nutrient addition? Plant Soil 313(1):141-153

Restrepo JI, Montoya AM, Obeysekera J (1998) A Wetland Simulation Module for the MODFLOW Ground Water Model. Groundwater 36(5):764-770

Rockström J, Falkenmark M (2015) Agriculture: increase water harvesting in Africa. Nature 519(7543):283-285

Rongoei PJK, Kariuki S (2019) Implications of Papyrus (*Cyperus papyrus* L.) Biomass Harvesting on Nutrient Regulation in Nyando Floodplain Wetland, Lake Victoria, Kenya. Open Journal of Ecology 9: 443-457

Rongoei PJK, Kipkemboi J, Kariuki ST, van Dam AA (2014) Effects of water depth and livelihood activities on plant species composition and diversity in Nyando floodplain wetland, Kenya. Wetl Ecol Manage, DOI 10.1007/s11273-013-9313-7

Rongoei PJK, Kipkemboi J, Okeyo-Owuor JB, van Dam AA (2013) Ecosystem services and drivers of change in Nyando floodplain wetland, Kenya. 7: 274-291

Rongoei PJK, Outa NO (2016) *Cyperus papyrus* L. Growth Rate and Mortality in Relation to Water Quantity, Quality and Soil Characteristics in Nyando Floodplain Wetland, Kenya. Open Journal of Ecology 6: 714-735

Russi D, ten Brink P, Farmer A, Badura Y, Coates D, Förster J, Kumar R, Davidson N (2013) The Economics of Ecosystems and Biodiversity for Water and Wetlands. IEEP, London and Brussels; Ramsar Secretariat, Gland

Rykiel Jr EL (1996) Testing ecological models: the meaning of validation. Ecol Model 90:229-244

Saltelli S, Tarantola S, Campolongo F (2000) Sensitivity analysis as an ingredient of modelling. Stat Sci 15(4):377-395

Saunders MJ, Jones MB, Kansiime F (2007) Carbon and water cycles in tropical papyrus wetlands. Wetl Ecol Manag 15(6):489-498

Saunders MJ, Kansiime F, Jones MB (2014) Reviewing the carbon cycle dynamics and carbon sequestration potential of *Cyperus papyrus* L. wetlands in tropical Africa. Wetl Ecol Manag 22:143-155

Savenije HHG (1995) New definitions for moisture recycling and the relationship with land-use changes in the Sahel. Journal of Hydrology 167(1):57-78

Schindewolf M, Schmidt J (2012) Parameterization of the EROSION 2D/3D soil erosion model using a small-scale rainfall simulator and upstream runoff simulation. CATENA 91:47-55

Schoumans OF, Bouraoui F, Kabbe C, Oenema O, van Dijk KC (2015) Phosphorus management in Europe in a changing world. Ambio 44: 180–192

Schuyt KD (2005) Economic consequences of wetland degradation for local populations in Africa. Ecol Econ 53:177-190

Sepúlveda R, Leiva AM, Vidal G (2020) Performance of Cyperus papyrus in constructed wetland mesocosms under different levels of salinity. Ecological Engineering 151: p.105820

Serag MS (2003) Ecology and biomass production of *Cyperus papyrus* L. on the Nile bank at Damietta, Egypt. J Mediterranean Ecol 4(3-4):15-24

Sheather SJ (2009) A modern approach to regression with R. Springer Science + Business Media, New York

Silvius MJ, Oneka M, Verhagen A (2000) Wetlands: lifeline for people at the edge. Physics and Chemistry of the Earth, Part B 25: 645–652

Sjögersten S, Black CR, Evers S, Hoyos-Santillan J, Wright EL, Turner BL (2014) Tropical wetlands: A missing link in the global carbon cycle? Global Biogeochem. Cycles 28, 1371–1386

Sklar FH, Costanza R, Day Jr JW (1985) Dynamic spatial simulation modeling of coastal wetland habitat succession. Ecological Modelling 29(1-4):261-281

Snyder JM, Rejmánková E (2015) Macrophyte root and rhizome decay: the impact of nutrient enrichment and the use of live versus dead tissue in decomposition studies. Biogeochemistry 124(1):45-59

Soetaert K, Hoffmann M, Meire P, Starink M, van Oevelen D, van Regenmortel S, Cox T (2004) Modeling growth and carbon allocation in two reed beds (*Phragmites australis*) in the Scheldt estuary. Aquatic Botany 79(3):211-234

Soissons LM, van Katwijk MM, Li B, Han Q, Ysebaert T, Herman PMJ, Bouma TJ (2019) Ecosystem engineering creates a new path to resilience in plants with contrasting growth strategies. Oecologia 191: 105-1024

Sollie S, Janse JH, Mooij WM, Coops H, Verhoeven JTA (2008) The contribution of marsh zones to water quality in Dutch shallow lakes: a modeling study. Environmental Management 42(6):1002-1016

Sorrell BK, Mendelssohn IA, McKee KL, Woods RA (2000) Ecophysiology of Wetland Plant Roots: A Modelling Comparison of Aeration in Relation to Species Distribution. Annals of Botany 86(3):675-685

St-Hilaire F, Wu J, Roulet NT, Frolking S, Lafleur PM, Humphreys ER, Arora V (2008) McGill Wetland Model: evaluation of a peatland carbon simulator developed for global assessments. Biogeosciences Discussions 5(2):1689-1725

Stuip MAM, Baker CJ, Oosterberg W (2002) The Socio-economics of Wetlands, Wetlands International and RIZA, The Netherlands

Su M, Stolte WJ, van der Kamp G (2000) Modelling Canadian prairie wetland hydrology using a semi-distributed streamflow model. Hydrological Processes 14(14):2405-2422

Sun G, Li C. Trettin CC, Lu J, McNulty SG (2006) Simulating the biogeochemical cycles in cypress wetland-pine upland ecosystems at a landscape scale with the wetland-DNDC model. In Hydrology and Management of Forested Wetlands, Proceedings of the International Conference, April 8-12, 2006, New Bern, North Carolina (p. 32). American Society of Agricultural and Biological Engineers

Tanaka N, Asaeda T, Hasegawa A, Tanimoto K (2004) Modelling of the long-term competition between *Typha angustifolia* and *Typha latifolia* in shallow water — effects of eutrophication, latitude and initial advantage of belowground organs. Aquat Bot. 79(4):295-310

TEEB (2010) The Economics of Ecosystems and Biodiversity for Local and Regional Policy Makers. Availabe at: http://www.teebweb.org/publications/teeb-study-reports/local-and-regional/

Temmerman S, Govers G, Meire Px, Wartel S (2003) Modelling long-term tidal marsh growth under changing tidal conditions and suspended sediment concentrations, Scheldt estuary, Belgium. Marine Geology 193(1-2):151-169

Terer T, Muasya AM, Dahdouh-Guebas F, Ndiritu GG, Triest L (2012a) Integrating local ecological knowledge and management practices of an isolated semi-arid papyrus swamp (Loboi, Kenya) into a wider conservation framework. J Environ Manage 93(1):71-84

Terer T, Triest L, Muasya AM (2012b) Effects of harvesting *Cyperus papyrus* in undisturbed wetland, Lake Naivasha, Kenya. Hydrobiologia 680(1):135-148

Terer T, Muasya AM, Higgins S, Gaudet JJ, Triest L (2014) Importance of seedling recruitment for regeneration and maintaining genetic diversity of *Cyperus papyrus* during drawdown in Lake Naivasha, Kenya. Aquatic Botany 116: 93-102

The United Republic of Tanzania (2013) 2012 Population and housing census, population Distribution by Administrative Areas. National Bureau of Statistics, Ministry of Finance, Dar es Salaam & Office of Chief Government Statistician President's Office, Finance, Economy and Development Planning, Zanzibar

Thompson K, Hamilton AC (1983) Peatlands of the African continent, Ecosystems of the World, Mires: Swamp, Bog, Fen and Moor, Regional Studies, 4BA. Gore JP ed, 331–373, Elsevier, New York, USA

Thompson K, Shewry PR, Woolhouse HW (1979) Papyrus swamp development in the Upemba Basin, Zaïre: studies of population structure in *Cyperus papyrus* stands. Bot J Linn Soc 78(4):299-316

Tickner D, Opperman JJ, Abell R, Acreman M, Arthington AH, Bunn SE, Cooke SJ, Dalton J, Darwall W, Edwards G, Harrison IAN, Hughes K, Jones TIM, Leclère D, Lynch AJ, Leonard P, Mcclain ME, Muruven D, Olden JD, Ormerod SJ, Robinson J, Tharme RE, Thieme M, Tockner K, Wright M, Young L (2020) Bending the Curve of Global Freshwater Biodiversity Loss: An Emergency Recovery Plan Forum. BioScience 70: 330–342

Toner M, Keddy P (1997) River hydrology and riparian wetlands: a predictive model for ecological assembly. Ecological applications 7(1):236-246

UN (2015) World Population Prospects: The 2015 Revision, Key Findings and Advance Tables. Working Paper No. ESA/P/WP.241. UN Department of Economic and Social Affairs, Population Division, New York

UNCCD (2017) The Global Land Outlook, first edition. Secretariat of the United Nations Convention to Combat Desertification, Bonn

UNDP (2012) Africa human development report 2012: Towards a food secure future. United Nations Publications, New York

Van Asselen S, Verburg PH, Vermaat JE, Janse JH (2013) Drivers of wetland conversion: a global meta-analysis. PLoS ONE 8: e81292

Van Dam AA, Dardona A, Kelderman P, Kansiime F (2007) A simulation model for nitrogen retention in a papyrus wetland near Lake Victoria, Uganda (East Africa). Wetland Ecology and Management 15, 469-480

Van Dam AA, Kipkemboi J (2018) Sustainable use of papyrus (Lake Victoria wetlands, Kenya). In: Davidson NC, Middleton B, McInnes R, Everard M, Irvine K, van Dam AA, Finlayson CM, editors. The Wetland Book Vol. 1: Wetland structure and function, management, and methods. Dordrecht (The Netherlands): Springer. https://doi.org/10.1007/978-90-481-9659-3_207

Van Dam AA, Kipkemboi J, Mazvimavi D, Irvine K (2014) A synthesis of past, current and future research for protection and management of papyrus (*Cyperus papyrus* L.) wetlands in Africa. Wetl Ecol Manag 22:99-114

Van Dam AA, Kipkemboi J, Rahman MM, Gettel GM (2013) Linking hydrology, ecosystem function, and livelihood outcomes in African papyrus wetlands using a Bayesian network model. Wetlands 33: 381-397

Van Dam AA, Kipkemboi J, Zaal AM, Okeyo-Owuor JB (2011) The ecology of livelihoods in East African papyrus wetlands (ECOLIVE). Project Update. Reviews in Environmental Science and Biotechnology 10, 291-300. DOI : 10.1007/s11157-011-9255-6

Van der Peijl MJ, Verhoeven JTA (1999) A model of carbon, nitrogen and phosphorus dynamics and their interactions in river marginal wetlands. Ecological Modelling 118, 95-130 Viner AB (1982) Nitrogen fixation and denitrification in sediments of two Kenyan lakes. Biotropica 14:91-98

Van der Peijl MJ, Verhoeven JTA (2000) Carbon, nitrogen and phosphorus cycling in river marginal wetlands; a model examination of landscape geochemical flows. Biogeochemistry 50(1):45-71

Van Horssen PW, Schot PP, Barendregt A (1999) A GIS-based plant prediction model for wetland ecosystems. Landscape Ecology 14(3)253-265

Van Huissteden J, van den Bos R, Alvarez IM (2006) Modelling the effect of water-table management on CO 2 and CH 4 fluxes from peat soils. Netherlands Journal of Geosciences 85(1):3-18

Verhoeven JTA, Setter TL (2010) Agricultural use of wetlands: opportunities and limitations. Annals of Botany 105: 155 –163

Villa F, Bagstad KJ, Voigt B, Johnson GW, Portela R, Honzák M, Batker D (2014) A Methodology for Adaptable and Robust Ecosystem Services Assessment. PLOS ONE 9(3):e91001

Viner AB (1982) Nitrogen fixation and denitrification in sediments of two Kenyan lakes. Biotropica 14:91-98

Walter BP, Heimann M (2000) A process-based, climate-sensitive model to derive methane emissions from natural wetlands: Application to five wetland sites, sensitivity to model parameters, and climate. Global Biogeochemical Cycles 14(3):745-765

Wang Q, Jørgensen SE, Lu J, Nielsen SN, Zhang J (2013) A model of vegetation dynamics of *Spartina alterniflora* and *Phragmites australis* in an expanding estuarine wetland: Biological interactions and sedimentary effects. Ecological Modelling 250:195-204

Wang N, Mitsch WJ (2000) A detailed ecosystem model of phosphorus dynamics in created riparian wetlands. Ecological Modelling 126(2):101-130

Weiner J (2004) Allocation, plasticity and allometry in plants. Perspectives in Plant Ecology, Evolution and Systematics 6: 207-215

Wood AP (2013) People centred wetland management. In: Wetland Management and Sustainable Livelihoods in Africa. Wood AP, Dixon A, McCartney M (eds). Taylor & Francis, pp. 142. ISBN 9781849714112

Wu J, Roulet NT, Nilsson M, Lafleur P, Humphreys E (2012) Simulating the carbon cycling of northern peatlands using a land surface scheme coupled to a wetland carbon model (CLASS3W-MWM). Atmosphere-ocean 50(4):487-506

Xu Z, Yang Z, Yin X, Cai Y, Sun T (2016) Hydrological management for improving nutrient assimilative capacity in plant-dominated wetlands: A modelling approach. J Ecol Manage 177:84-92

Yang G, Best EPH (2015) Spatial optimization of watershed management practices for nitrogen load reduction using a modeling-optimization framework. Journal of Environmental Management 161:252-260

Ye F, Chen Q, Blanckaert K, Ma J (2013) Riparian vegetation dynamics: insight provided by a process-based model, a statistical model and field data. Ecohydrology 6(4):567-585

Yu Z, Loisel J, Brosseau DP, Beilman DW, Hunt SJ (2010) Global peatland dynamics since the Last Glacial Maximum. Geophysical Research Letters 37(13): L13402

Zemlin R, Kühl H, Kohl J-G (2000) Effects of seasonal temperature on shoot growth dynamics and shoot morphology of common reed (*Phragmites australis*). Wetlands Ecology and Management 8: 447-457

Zhang Y, Miao Z, Bognar J, Lathrop RG (2011) Landscape scale modeling of the potential effect of groundwater-level declines on forested wetlands in the New Jersey Pinelands. Wetlands 31(6):1131-1142

Zhang Y, Li C, Trettin CC, Li H, Sun G (2002) An integrated model of soil, hydrology, and vegetation for carbon dynamics in wetland ecosystems. Global Biogeochemical Cycles 16(4):9-1-9-17

Zhou L, Zhou G (2009) Measurement and modelling of evapotranspiration over a reed (*Phragmites australis*) marsh in Northeast China. Journal of Hydrology 372(1):41-47

Zhu Q, Peng C, Chen H, Fang X, Liu J, Jiang H, Yang Y, Yang G (2015) Estimating global natural wetland methane emissions using process modelling: spatio-temporal patterns and contributions to atmospheric methane fluctuations. Global Ecol Biogeogr 24(8):959-972

LIST OF ACRONYMS[5]

AFDW	Ash-free dry weight
AGB	Aboveground biomass
ANOVA	Analysis of variance
APHA	American Public Health Association
BGB	Belowground biomass
C	Carbon
DW	Dry weight
FAO	Food and Agriculture Organization of the United Nations
GIS	Geographical Information System
IPBES	Intergovernmental Science-Policy Platform on Biodiversity and Ecosystem Services
MEA	Millennium Ecosystem Assessment
N	Nitrogen
OAT	One-at-a-atime
OECD	Organization for Economic Cooperation and Development
P	Phosphorus
SDG	Sustainable Development Goal
TEEB	The Economics of Ecosystems and Biodiversity

---

[5] For complete lists of model variables, see Appendices 4.1 (p. 67), 5.1 (p. 119) and 5.2 (p. 120)

LIST OF TABLES

LIST OF FIGURES

ABOUT THE AUTHOR

Edwin Hes obtained an MSc in Environmental Science and Technology from Wageningen University in 1998. From 1999 to 2002 he worked for the Dutch Ministry of Economic Affairs as a public consultant on European research grants. In 2002, he started at IHE Delft Institute for Water Education and worked for the Liaison Office until 2007. Later that year, he started working as a Lecturer in Environmental Systems Analysis. His main research interest during the past years was modelling nutrient flows in wetland ecosystems and how these are influenced by human activities, especially in papyrus wetlands in East Africa. In general, he is interested in modelling and quantifying wetland ecosystem functions and services at local and global levels. Being a generalist, he enjoys to contribute to research across disciplines from engineering (constructed wetlands for wastewater treatment) to governance (related to wetland ecosystems) and economics (valuation of ecosystem services). More recently he developed an interest in transdisciplinary research and specifically in questions related to the influence of the dominance of Western scientific concepts and approaches on global conservation.

At IHE Delft, he has been coordinating the Environmental Science MSc programme from 2012 to 2016 and the joint degree programme in Limnology and Wetland Management since 2012. Hes has been involved with teaching and training activities in the fields of Environmental Systems Analysis, Environmental Modelling and Wetland Management for 15 years.

His activities are often in collaboration with colleagues from partner organizations, most notably from Mexico (LANCIS-UNAM), Kenya (Egerton University), Austria (BOKU) and the Netherlands (WUR and PBL). Since 2007, Hes has worked in research and capacity building projects in more than 20 countries in Asia, Africa and Latin America with a focus on sustainable use of wetlands.

Publications

Hes EMA, Yatoi R, Laisser SL, Feyissa AK, Irvine K, Kipkemboi J, van Dam AA (2020) The effect of seasonal flooding and livelihood activities on retention of nitrogen and phosphorus in Cyperus papyrus wetlands, the role of aboveground biomass. Hydrobiologia, https://doi.org/10.1007/s10750-021-04629-3

Hes EMA, van Dam AA (2019) Modelling nitrogen and phosphorus cycling in Cyperus papyrus dominated natural wetlands. Environmental Modelling and Software 122, 104531. https://doi.org/10.1016/j.envsoft.2019.104531

Janse JH, van Dam AA, Hes EMA, de Klein JJM, Finlayson CM, Janssen ABG, van Wijk D, Mooij WM, Verhoeven JTA (2019) Towards a global model for wetlands ecosystem services. Current Opinion in Environmental Sustainability 36, 11-19. https://doi.org/10.1016/j.cosust.2018.09.002

Hes EMA, Niu R, van Dam AA (2014) A simulation model for nitrogen cycling in natural rooted papyrus wetlands in East Africa. Wetlands Ecology and Management 22, 157-176. doi:10.1007/s11273-014-9336-8

Gope ET, Sass-Klaassen UGW, Irvine K, Beevers L, Hes EMA (2015) Effects of flow alteration on Apple-ring Acacia (Faidherbia albida) stands, Middle Zambezi floodplains, Zimbabwe. Ecohydrology 8 (5), 922-934

Ncube S, Beevers L, Hes EMA (2013) The interactions of the flow regime and the terrestrial ecology of the Mana Floodplains in the Middle Zambezi river basin. Ecohydrology 6(4), 554-566

Guzman-Ruiz A, Schwartz KH, Hes EMA (2011) Shifting Governance Modes in Wetland Management: A Case Study of two Wetlands in Bogotá, Colombia. Environment and Planning C 29, 990-1003

Babu MA, Hes EMA, van der Steen NP, Hooijmans CM, Gijzen HJ (2010) Nitrification rates of algal-bacterial biofilms in wastewater stabilization ponds under light and dark conditions. Ecological Engineering 36 (12), 1741-1746

Rickwood CJ, Hes EMA, Al-Zu'bi Y, Dubé MG (2010) Overview of limitations, and proposals for improvement, in education and capacity building of Ecohydrology. Ecohydrology and Hydrobiology 10 (1), 45-59

Schouten MAC, Hes EMA, Hoko Z (2009) Innovative Practices in the African Water Supply and Sanitation Sector. SUN MeDIA, Stellenbosch, South Africa

Netherlands Research School for the
Socio-Economic and Natural Sciences of the Environment

D I P L O M A

for specialised PhD training

The Netherlands research school for the
Socio-Economic and Natural Sciences of the Environment
(SENSE) declares that

Edwin Marinus Andries Hes

born on 26 November 1973 in Meppel, The Netherlands

has successfully fulfilled all requirements of the
educational PhD programme of SENSE.

Wageningen, 1 October 2021

Chair of the SENSE board

Prof. dr. Martin Wassen

The SENSE Director

Prof. Philipp Pattberg

The SENSE Research School has been accredited by the Royal Netherlands Academy of Arts and Sciences (KNAW)

K O N I N K L I J K E   N E D E R L A N D S E
A K A D E M I E   V A N   W E T E N S C H A P P E N

The SENSE Research School declares that Edwin Marinus Andries Hes has successfully fulfilled all requirements of the educational PhD programme of SENSE with a work load of 47.4 EC, including the following activities:

SENSE PhD Courses

o   Research in context activity: 'Multidisciplinary course development on Wetlands for Livelihoods and Conservation' (2012)

Other PhD and Advanced MSc Courses

o   Environmental modelling, dynamic simulation modelling with Stella, IHE Delft (2010)

Selection of Management and Didactic Skills Training

o   Supervising MSc students with thesis (2010-2021)
o   Course coordinator Wetlands for Livelihoods and Conservation, IHE Delft (2012-2021)
o   Course leader Wetland Management for Indonesian Professionals, Biotrop Indonesia (2011)
o   Coordinator of the MSc programme Environmental Science IHE Delft (2012-2016)
o   Coordinator of the joint Master of Science programme in Limnology and Wetland Management (LWM) with BOKU University, Vienna, Austria; Egerton University, Njoro, Kenya and IHE Delft (2012-2021)
o   Course leader on regional workshop on working with Ecosystem Services to understand and manage wetlands, L'institut National De L'eau (Ine) Au Benin (2014)
o   Guest lecturer on socio-ecological systems, Universidad Nacional Autónoma de México (2019-2021)

Selection of Oral Presentations

o   *The Papyrus Simulator, Modelling Nitrogen and Phosphorus Cycling in Natural Rooted Papyrus Wetlands*. 10th INTECOL International Wetlands Conference, 19-24 September 2016, Changshu, PR China
o   *A simulation model for nitrogen and phosphorus retention in seasonally flooded and permanently flooded wetlands in East Africa*. 9th INTECOL International Wetlands Conference, 3-8 June 2012, Orlando, United States of America
o   *A simulation model for nitrogen retention in seasonally flooded and permanently flooded wetlands in East Africa*, Society of Wetland Scientists, 3-8 July 2011, Prague, Czech Republic
o   *IWRM and Environmental Flows principles to support management of the Gulf of Mexico Large Marine Ecosystem*. Regional International Forum: "From the rivers to the Gulf of Mexico, towards and ecosystem management approach", 27-28 September 2010, Mexico City, Mexico

SENSE coordinator PhD education

Dr. ir. Peter Vermeulen